Governing Uncertainty

Environmental Regulation in the Age of Nanotechnology

Edited by

Christopher J. Bosso

RFF PRESS
RESOURCES FOR THE FUTURE

Washington, DC • London

First published in 2010 by RFF Press, an imprint of Earthscan

Copyright © Earthscan 2010

Earthscan LLC, 1616 P Street, NW, Washington, DC 20036, USA
Earthscan Ltd, Dunstan House, 14a St Cross Street, London EC1N 8XA, UK
Earthscan publishes in association with the International Institute for Environment and Development

For more information on RFF Press and Earthscan publications, see www.rffpress.org and www.earthscan.co.uk or write to earthinfo@earthscan.co.uk

ISBN: 978-1-84407-807-3 (hardback)
ISBN: 978-1-84407-808-0 (paperback)

Copyedited by Joyce Bond
Typeset by JS Typesetting Ltd, Porthcawl, Mid Glamorgan
Cover design by Ellen A. Davey
Cover image from istock International Inc.

Library of Congress Cataloging-in-Publication Data

Bosso, Christopher J. (Christopher John), 1956-
    Governing uncertainty : environmental regulation in the age of nanotechnology / edited by Chistopher J. Bosso.
        p. cm.
    Includes bibliographical references and index.
    ISBN 978-1-933115-80-1 (hardback) -- ISBN 978-1-933115-79-5 (pbk.)
    1. Nanotechnology. 2. Nanotechnology--Environmental aspects. 3. Nanoparticles--Environmental aspects.
    T174.7.G68 2010
    620'.5--dc22
                            2010001529

A catalog record for this book is available from the British Library.

Mixed Sources
Product group from well-managed forests and other controlled sources
www.fsc.org   Cert no. SGS-COC-004946
© 1996 Forest Stewardship Council
FSC

About Resources for the Future *and* RFF Press

Resources for the Future (RFF) improves environmental and natural resource policymaking worldwide through independent social science research of the highest caliber. Founded in 1952, RFF pioneered the application of economics as a tool for developing more effective policy about the use and conservation of natural resources. Its scholars continue to employ social science methods to analyze critical issues concerning pollution control, energy policy, land and water use, hazardous waste, climate change, biodiversity, and the environmental challenges of developing countries.

RFF Press supports the mission of RFF by publishing book-length works that present a broad range of approaches to the study of natural resources and the environment. Its authors and editors include RFF staff, researchers from the larger academic and policy communities, and journalists. Audiences for publications by RFF Press include all of the participants in the policymaking process—scholars, the media, advocacy groups, NGOs, professionals in business and government, and the public.

CONTENTS

ACKNOWLEDGMENTS

This project is supported by a National Science Foundation award, Nano-technology in the Public Interest: Regulatory Challenges, Capacity, and Policy Recommendations (SES-0609078). The views expressed here are solely those of the authors. Thanks go to Carolyn Oakes and Matthew Botti of the NSF Center for High-rate Nanomanufacturing (EEC-425826), at Northeastern University, and Ahmed Busnaina, director, for their help in setting up the research workshops that led to this volume, and James Stellar, dean of arts and sciences, and Srinivas Sridhar, vice provost for research, for additional financial support. We also thank our able research assistants, Hans Eijmberts, Mark Griffin, Lindsay Dahlben, Katrina McCarty, Caitlin McAllister, and Zeynep Ok. Special thanks also go to colleague Ronald Sandler, whose insights sharpened our thinking throughout this volume. Finally, our thanks to RFF Press publisher Don Reisman, the RFF Press staff, and the anonymous reviewers who made this volume all the better.

CONTRIBUTORS

Christopher J. Bosso (editor) is professor of public policy and associate dean of the School of Public Policy and Urban Affairs at Northeastern University. His recent book *Environment, Inc.: From Grassroots to Beltway* received the 2006 Caldwell Award from the American Political Science Association. He is principal investigator on the National Science Foundation Nanotechnology Interdisciplinary Research Team, Nanotechnology in the Public Interest, and director of the Nanotechnology and Society Research Group (www. nsrg.neu.edu), which conducts research on regulatory and environmental dimensions of nanotechnology.

Cary Coglianese is associate dean for academic affairs and Edward B. Shils Professor of Law at the University of Pennsylvania, where he is also professor of political science and the founding director of the Penn Program on Regulation. He studies regulation and regulatory processes, the evaluation of alternative regulatory strategies, and business–government relations in regulatory policymaking. His scholarly works include *Regulating from the Inside: Can Environmental Management Systems Achieve Policy Goals?* and *Leveraging the Private Sector: Management-Based Strategies for Improving Environmental Performance, and Regulation and Regulatory Processes.*

J. Clarence (Terry) Davies is a senior advisor to the Project on Emerging Nanotechnologies and a senior fellow at Resources for the Future. He chaired the National Academy of Sciences Committee on Decision Making for Regulating Chemicals in Environment. While serving as a consultant to the President's Advisory Council on Executive Organization, he coauthored the reorganization plan that created the U.S. Environmental Protection Agency (EPA).

Marc Allen Eisner is the Henry Merritt Wriston Chair of Public Policy and professor of government at Wesleyan University. He is author or coauthor of six books, on topics ranging from the changing role of economic analysis in antitrust policy to the impact of World War I mobilization on interwar state building. His most recent book is *Governing the Environment: The Transformation of Environmental Regulation*. His current research focuses on the integration of public regulation and association- and standards-based self-regulation in environmental protection; and the challenges of designing regulatory institutions.

Jacqueline A. Isaacs is professor of mechanical and industrial engineering and associate director of the Center for High-rate Nanomanufacturing (CHN) at Northeastern University. Her research focuses on economic and environmental assessment of manufacturing and the societal implications of technology. She also has received National Science Foundation (NSF) funding to develop an interactive educational game called Shortfall!—which encourages students to make decisions based on technological, economic, and environmental trade-offs within a production facility.

W. D. Kay is associate professor of political science at Northeastern University, where he specializes in organization theory as well as science and technology policy, with a particular focus on the politics of innovation. He is author of two scholarly books on the U.S. space program, *Can Democracies Fly in Space?* and *Defining NASA*, and is leading a nanotechnology interdisciplinary research team project on historical examples of government responses to emerging technologies.

Marc Landy is professor of political science at Boston College. He is author or coauthor (with Sidney M. Milkis) of *Presidential Greatness, The Environmental Protection Agency from Nixon to Clinton: Asking the Wrong Questions*, and a textbook, *American Government: Balancing Democracy and Rights*. His edited books include *Creating Competitive Markets: The Politics and Economics of Regulatory Reform, Seeking the Center: Politics and Policymaking at the New Century*, and *The New Politics of Public Policy*.

Sean T. O'Donnell is a historian of science and postdoctoral researcher at Harvard University, where he collaborates on an NSF-funded project studying the use of scientific evidence in state and federal courts. He is also a lecturer in medical sociology at Tufts University. His work approaches the intersection of law and science from social, cultural, and historical perspectives. His Harvard dissertation, Courting Science, Binding Truth: A Social History of *Frye v. U.S.*, recasts the history of the legal standard for the admissibility of scientific evidence amid the racial politics of the Washington, D.C., courts during the pre-Civil Rights era.

Barry G. Rabe is professor at the University of Michigan, with appointments in the Ford School of Public Policy, School of Natural Resources and the Environment, and Program in the Environment, and a nonresident senior fellow at the Brookings Institution. He is author of four books, including *Statehouse and Greenhouse*, which received the Caldwell Award from the American Political Science Association in 2005. In 2006, he received a Climate Protection Award from the U.S. Environmental Protection Agency. He currently directs research on climate governance and cross-border environmental policy in North America.

FOREWORD: NANOTECHNOLOGY, RISK, AND GOVERNANCE

J. Clarence (Terry) Davis

The authors of *Governing Uncertainty* have used the example of nanotechnology both as an analytical microscope to better understand the management of environmental health and safety problems and as an analytical telescope to foresee the challenges that lie ahead. They have skillfully examined the nexus of technological development and social institutions, a nexus where much of the world's future will be played out.

THE PROMISE OF NANOTECHNOLOGY

Nanotechnology is the science and application of manipulating matter at the scale of individual atoms and molecules. All natural processes, from the growth of human embryos to plant photosynthesis, operate in this way, but only recently have we developed the tools that allow us to build and analyze things at the molecular level. For the first time in human history, we are close to being able to manipulate the basic forms of all things, living and inanimate, taking them apart and putting them together in almost any way the mind can imagine. The world of the future will be defined by how we use this mastery.

The benefits of nanotechnology, both current and future, are hard to exaggerate. Nanotechnology is used now to make car bodies stronger and lighter, to make batteries and solar panels more efficient, to make glass that never needs cleaning and neckties that are stain-proof, and to deliver medicines to individual cells in the body. In the future, assuming that the technology is not impeded by public opposition, "nano" will bring us water desalination at a fraction of the current cost, materials that can make objects invisible, revolutionary new types of computers, and medicines that will cure many major diseases.

The technology also has potential risks, and no nation—including the United States—has the oversight policies and institutions needed to deal with these risks.

IGNORANCE DOES NOT LEAD TO BLISS

We know very little about the risks of nanotechnology. To date, there are no documented instances of consumers or users being harmed by the technology or its applications. Nevertheless, we know enough about the technology and have enough experience with other technologies to confidently predict that some kinds of problems will arise and other kinds of problems should be guarded against. Here I will discuss three facets of risk: risk to human health and the natural environment, concerns about social values, and the importance of perceived risk.

Risk to Health and the Environment

There is a dearth of information about the health and environmental risks of nanomaterials because the technology is relatively new and insufficient resources have been devoted to understanding its risks. Of the $1.5 billion the U.S. government is spending annually on nano research and development, less than 3 percent is for research to identify health and environmental risks, and even this small amount is not being spent in accordance with any well-formulated strategy.

But we have several reasons to be concerned about nano's health and environmental effects. First, nanomaterials commonly behave differently than materials of ordinary size, often following different laws of physics, chemistry, and biology. For example, aluminum is harmless when used in soft-drink cans, but nanoscale aluminum is so explosive that it is being considered as bomb-making material by the military. The differences between nanomaterials and ordinary materials mean that much of our existing knowledge about risks is not applicable to nanotechnology.

Second, one of the defining characteristics of nanomaterials is their very large surface area relative to their mass. It is on the surface of materials that chemical and biological reactions take place, so one would expect nanomaterials to be much more reactive than bulk materials. This is an advantage in many nano applications but can also be a potential hazard.

Third, the small size of nanomaterials means that they can get to places ordinary materials cannot. For example, there is some evidence that nano-materials can penetrate the blood-brain barrier. This could be an advantage for delivering medications but could be a serious danger if certain types of materials were inhaled or ingested.

A fourth reason for concern is that, based on past experience, it would be extraordinary if nanomaterials did not pose potential health and environ-mental problems. Our experience with bulk chemicals has taught us the necessity of oversight, and nanomaterials are, indeed, chemicals.

The results of nanomaterial toxicity tests using laboratory animals have been inconclusive to date but give cause for concern. They show that even small differences in the characteristics of a nanomaterial, such as its shape or the electrical charge at its surface, can make a big difference in its chemical and biological behavior. So testing done on substance A may not identify the risks of substance B, even though the two substances may seem almost identical. The most worrisome test results have shown that certain types of carbon nanotubes (a common form of nanomaterial), when inhaled by laboratory animals, produce the same type of precancerous lesions as asbestos. Other tests have indicated that some nanomaterials may damage DNA or certain types of cells. As more testing is done, these results will become more or less certain, and other effects are likely to be identified.

Social Risks

If one defines risk as the possibility of an adverse consequence, health and environmental risks are not the only kinds a technology may pose. People are often concerned about a technology being used in a way that conflicts with some deeply held value, and some uses of nano may create such conflict.

A 2004 study by Grove-White and others compared the issues in the controversy over biotechnology to those that might be expected in relation to nanotechnology. Their findings showed potentially strong similarities, including concerns about

> global drives towards new forms of proprietary knowledge; shifting patterns of ownership and control in the food chain; issues of corporate responsibility and corporate closeness to governments; intensifying relationships of science and scientists to the worlds of power and commerce; unease about hubristic approaches to limits in human understanding; and conflicting interpretations of what might be meant by sustainable development.

As the authors pointed out, these kinds of concerns cannot be accommodated within a framework of risk assessment of individual nanotechnology products.

Nano is also likely to raise a number of ethical questions that cannot be addressed within the usual risk assessment framework. If nanotechnology can be used to improve the functioning of the human brain, should it be used that way? And if so, for whose brains? If nanoscale materials are incorporated in foods to improve nutrition, shelf life, or taste, should the food have to be labeled to show that nano has been used? Synthetic biology, which is becoming increasingly merged with nanotechnology, is a new area of biological research that combines science and engineering in order to design and build, or synthesize, novel biological functions and systems. If synthetic biology can create new life forms, should it be allowed to do so? These and many other issues likely to be raised by nanotechnology in the not-too-distant future may pose potential risks to values.

Perceived Risk

The greatest threat to the development and application of nano may not be any actual documented risk, but rather a perception that the technology is risky and dangerous. Such views are produced by an amalgam of real risks, people's cultural orientation, information disseminated about the technology, perceptions of the adequacy of safeguards against risk, and probably other factors.

Because nanotechnology is new, invisible, and hard to explain in ordinary language, it lends itself to nonrational opinions. Polls show a large majority of people have little or no knowledge of the technology, but this is no bar to many of those polled having strong opinions on the subject. The experts, seeking to gain support for the technology or at least foster a more elevated debate about it, have supported public education about nano.

They have been cheered by studies showing that support for nano correlates with knowledge about the technology. Dan Kahan and others (2009) have shown, however, that the direction of causation is probably the reverse of what has been assumed. People who are culturally inclined to support new technologies are also more inclined to learn about the technologies. In experiments, providing additional information about nanotechnology to people whose cultural views were mistrustful of new technologies left the people more mistrustful of nano, even though the information was quite balanced.

It may be tempting to dismiss views based on a lack of information or on misinformation. Nevertheless, perceived risk is a real factor in people's behavior. If we want them to buy products containing nanomaterials or not support bans on nanotechnology research, we need to understand that perceived risks are at least as important as "real" risks.

NANO OVERSIGHT NEEDS

The U.S. regulatory system is not prepared to deal with nanotechnology or the other technological advances that lie ahead. In the near term, many of the changes needed to deal with nanotechnology are the same as those needed to remake the currently dysfunctional regulatory system. All four of the major environmental health and safety regulatory agencies—the U.S. Environmental Protection Agency (EPA), Food and Drug Administration (FDA), Occupational Safety and Health Administration (OSHA), and Consumer Product Safety Commission (CPSC)—are hobbled by antiquated and perverse laws and totally inadequate resources. These agencies need more money and more personnel with relevant expertise. And a significant increase in research on the risks posed by nanomaterials is needed as well.

Under their existing authority, the regulatory agencies could take numerous steps to improve oversight of nanoproducts and nanomaterials. To provide even minimally adequate oversight, however, legislative action is essential. The Toxic Substances Control Act (TSCA) is the only law that can regulate nanomaterials generally. It is a deeply flawed act that needs major overhauling, not just for nano, but for any type of chemical. The Federal Food, Drug, and Cosmetic Act regulates a variety of important uses of nanomaterials, such as in drugs and food, but it contains language that prevents oversight of two uses that involve high human exposure to nanomaterials: dietary supplements (vitamin pills, herbal remedies, and the like) and cosmetics.

In the longer term, the revolutionary scientific and technological innovations that are on the horizon will require totally different ways of dealing with potential risk. The future will be characterized by rapid scientific advancement and utilization of science, frequent product changes, technical complexity, and a variety of novel ethical, social, health, and environmental challenges. A regulatory system that takes two years to issue a rule cannot deal with an economy where product lines typically change every six months. A regulatory law focused on types of chemicals cannot deal with something like nanomaterials, where often the same chemical substance can have radically different effects depending on small changes in its shape or the method by which it is manufactured.

Many longer-term changes are needed. One of the most important would be the creation of a new Department of Environmental and Consumer Protection, which would incorporate six existing agencies: EPA, CPSC, OSHA, National Oceanic and Atmospheric Agency (NOAA), U.S. Geological Survey (USGS), and National Institute for Occupational Safety and Health (NIOSH). This new meta-agency would focus on science and monitoring, although it would have a strong oversight component. It would foster more integrated approaches, requiring new legislation. A clear need exists for a more integrated law focusing on dangerous products that would supersede such existing laws as TSCA and the Consumer Product Safety Act.

The United States is not prepared to deal with the challenges posed by twenty-first-century science and technology. Thinking and discussion about new approaches should start now. The future context for dealing with risk will be unlike anything we have known, and the policies of the past will not provide the protection we need. The authors of *Governing Uncertainty* have made an important contribution to getting the discussion started.

J. Clarence (Terry) Davies

FURTHER READING

Davies, J. Clarence. 2008. *Nanotechnology Oversight: An Agenda for the New Administration*. Washington, DC: Project on Emerging Nanotechnologies, Woodrow Wilson International Center for Scholars.

———. 2009. *Oversight of Next Generation Nanotechnology*. Washington, DC: Project on Emerging Nanotechnologies, Woodrow Wilson International Center for Scholars.

Grove-White, Robin, Matthew Kearnes, Paul Miller, Phil Macnaghten, James Wilsdon, and Brian Wynne. 2004. Bio-to-Nano? Working paper, Institute for Environment, Philosophy and Public Policy, Lancaster University and Demos.

Kahan, Dan M., et al. 2009. Cultural Cognition of the Risks and Benefits of Nano-technology. *Nature Nanotechnology* 4 (February): 87–90.

CHAPTER 1

POLICY CONSEQUENCES OF THE "NEXT INDUSTRIAL REVOLUTION"

Christopher J. Bosso

A simple query on the Google search engine in February 2009 listed nearly 10 million articles, reports, and websites on some facet of nanotechnology. Of those, more than 450,000 matched up the term "nanotechnology" with "revolution," and 11,000 with the even more precise phrase "next Industrial Revolution," echoing oft-cited claims about the transformative impacts of this enabling technology going back to when the Clinton White House kicked off the cross-agency National Nanotechnology Initiative (NNI) in January 2000 with its report *National Nanotechnology Initiative: Leading to the Next Industrial Revolution* (White House 2000). That language of "revolution" is rife in professional circles, perhaps even more than in the popular press. Note, for example, this depiction by the Society of Manufacturing Engineers:

> Nanotechnology promises to usher in the next Industrial Revolution and *replace our entire manufacturing base* with a new, radically precise, less expensive, and more flexible way of making products. These pervasive changes in manufacturing will leave virtually no product, process, or industry untouched. Nanotechnology has the potential to *disrupt entire industries* while leading to the creative destruction of current business models. (SME 2006; emphasis added)

1

Such heady predictions yield equally dramatic forecasts about nanotechnology's commercial potential, with widely bandied estimates of a global market ranging anywhere from $1 trillion to $3 trillion *a year* within a decade (Roco 2003). This, in turn, has prompted significant public- and private-sector investments to promote rapid development and commercialization of nanotechnology-based applications and products. Under the umbrella of the NNI, the U.S. federal government alone is spending nearly $1.5 billion annually in nanoscale science and engineering research and development (NNI 2008), a level of targeted federal spending on nondefense specific science and technology not seen since the heyday of the U.S. space program. Private-sector investments are estimated to equal, if not exceed, that figure, and state governments from New York and California to Ohio and Oklahoma are funding their own nanotechnology research and development programs to ensure their future economic competitiveness (EOP 2003). Similar initiatives are evident in any number of nations, notably but not exclusively South Korea, Germany, the United Kingdom, and China.

So nanotechnology, most everyone agrees, is the next big thing—the next Industrial Revolution. As a result, and not surprisingly, there are very high expectations, if not wildly inflated dreams, about the promise of nanotechnology to shape life in the twenty-first century in the same way that the petrochemical revolution defined a broad spectrum of basic materials and applications in second half of the twentieth century.

Yet enthusiasm about nanotechnology is partially tempered by recognition of the hurdles along the road to widespread commercialization, particularly the societal implications of this emerging technology, ranging from immediate concerns about the potential environmental and human health and safety effects of nanoscale materials on laboratory researchers and production workers to longer-term and more profound impacts of nanotechnologies directed toward enhancing the human body. Many of these concerns reflect previous experiences with the environmental and health side effects of the petrochemical revolution, such as the harm to bird populations caused by chemical pesticides (Bosso 1987) and concerns about systemic human health impacts of discarded pharmaceuticals in drinking water (Daughton and Ternes 1999). Indeed, in enacting the 21st Century Nanotechnology Research and Development Act in 2003, the U.S. Congress required that the National Science Foundation (NSF) in particular fund research into unspecified "ethical, legal, environmental, and other appropriate societal concerns" about nanotechnology and mandated that "public input and outreach to be integrated into the Program by the convening of regular and ongoing public discussions, through mechanisms such as citizens' panels, consensus conferences, and educational events, as appropriate" (U.S. Congress 2003).

Largely because Congress so mandated, concerns about societal implications are being addressed in tandem with technological development—if not exactly at the same level of funding—in contrast to the days when concerns about environmental and health impacts of new technologies often did not emerge until they became more clearly, and sometimes disastrously, manifest (Bosso 1987). In this regard, mounting concerns that too little is known about the possible environmental and human health impacts of nanomaterials prompted the NSF and U.S. Environmental Protection Agency (EPA) to jointly fund two new Centers for Environmental Implications of Nanotechnology—one at Duke University, the other at the University of California–Los Angeles—as part of a five-year, $25 million initiative dedicated to "understanding the interactions of nanomaterials with organisms, cellular constituents, metabolic networks and living tissues; understanding environmental exposure and bioaccumulation and their effects on living systems; and determining the biological impacts of nanomaterials dispersed in the environment." The NSF is clear that this initiative's value lies in "*reducing uncertainty* about the environmental health and safety implications of nanotechnology through research, education, public outreach and dissemination" (CEIN 2006; emphasis added).

Reducing uncertainty is critical on a number of fronts and for a great many potential stakeholders, and such basic research will yield important information on the fundamental properties of nanomaterials and risk factors they pose. Left unanswered, however, is exactly how any information obtained from this research will be used, by whom, and for what purposes. It is expected but unstated that any data obtained through toxicological and epidemiological research will help corporate and government leaders shape appropriate responses, but in what form and by whom is not clear. It is even less certain what role, if any, government will play in deciding the shape, application, or enforcement of any rules or procedures directed toward the prevention or swift remediation of environmental, health, and safety effects.

More troubling for the long term is a relative absence of open discussion about the capacity of government to use any information obtained. For all the expressed concern in government and industry circles about proactively addressing the possible environmental, health, and safety effects of nanotechnology—if only to obtain greater public acceptance of new technologies—to date relatively little attention has been paid to understanding the current institutional capacity and anticipated informational needs of the agencies and officials that researchers, firms, investors, and citizens expect to make critical decisions on a wide range of emerging nanotechnology applications.

And we know that such challenges will occur. After all, the Chemical Age that so defined the twentieth century, for all of its multiform benefits, ultimately generated tremendous direct and indirect environmental and human health consequences, ranging from toxicity and carcinogenicity in humans to impacts on wildlife populations. Public concerns about such effects—some of them misplaced—eventually prompted government at several levels to respond with an array of regulatory institutions, laws, and policy tools intended to obviate harmful environmental effects and protect human health (Bosso 1987). In some minds, nanotechnology, broadly conceived, may pose similar, if not greater, challenges.

Thus comes the core challenge for democratic government. Whatever our general ideological or partisan views about the "proper" role of government, most of us at some level do expect it to protect us from those risks we as individuals can neither comprehend nor control. As individuals, we may have some say about what we drink, eat, or even smoke, but we have little control over the basic technologies that underpin our daily lives, be they the genetic map of new grain hybrids; the materials that make up our electronics, modes of transportation, and buildings; or the core substances in our pharmaceutical and medical devices. So we enlist a third party—usually the government—to act on our behalf in ensuring that the stuff of everyday life does not cause us undue harm. And as citizens in the contemporary age, we usually expect government to be reasonably open and responsive to our concerns in devising any policy responses to those risks. As we have seen in other instances, such as the muddled government response to the outbreak of bovine spongiform encephalopathy, or "mad cow" disease, in the United Kingdom in the 1990s (Tarrow 2000), any apparent failure to respond to perceived risks in an effective, responsive, and transparent way undermines public trust in government itself. Citizens' loss of faith in government capacity or transparency, in turn, can undermine public receptivity to policymakers' claims that new technologies pose little risk. Based on even cautious estimates of nanotechnology's expected reach and impacts (discussed in Chapter 2), this basic challenge to government will only grow more acute.

ABOUT THIS BOOK

Questions about the fundamental capacity of government to balance a desire to foster technological innovation and economic growth with the protection of public health and the natural environment prompted the NSF to support a range of research efforts, one being the multiyear project on nanotechnology and government capacity out of which this volume grew (Bosso et al. 2006). The overall project evaluates existing government capacity—defined here as sufficiency in scientific expertise, legal authority, organizational design, and relevant regulatory frameworks—to address the societal and policy challenges posed by nanoscale innovations and products and, where appropriate, to make recommendations for building requisite capacity to address these challenges.

Such concerns are not simply reducible to more money, more people, or more time, and instead focus on the following core questions: What have we learned from decades of government understanding about, concern with, and response to environment, health, and safety effects generated by new technologies and their applications, and how do we apply those lessons to new technologies of as yet unknown properties and uncertain effects? How should democratic government approach inevitable questions of responding to the direct and indirect effects of new and potentially revolutionary technologies? How does government balance the imperatives of technological innovation and economic growth with citizen demands that it protect public health and the natural environment? And, of concern to students of democratic government in particular, who decides?

In an early assessment on the applicability of the federal Toxic Substances Control Act (TSCA) to nanoparticles, the Foresight and Governance Project at the Woodrow Wilson Center for International Scholars offered four conclusions with relevance to nanotechnology more broadly conceptualized:

- The unique properties inherent in nanotechnology will pose new challenges to existing regulatory structures and, in the process, create confusion within both industry and government about the nature and scope of regulation.
- Little attention has been paid to the adequacy of the current regulatory system to protect human health and the environment, or about possible alternatives to existing regulatory regimes.

- The absence of any conclusive understanding about the health risks of nano-based substances makes more urgent the need for attention to and a dialog on regulatory adequacy and needed changes.
- Misguided or poorly designed regulatory approaches could have enormous economic consequences. (Wardak 2003)

In sum, there are expressed concerns about existing federal and state government capacity to respond to, much less proactively address, the societal and policy challenges posed by nanoscale substances and innovations. Though it may not be clear how nanotechnology will affect, and be affected by, the regulatory landscape, no doubt these interactions will occur, with real consequences. Those interested in nanotechnology therefore have a keen interest in how these as-yet-uncertain dynamics will play out. Moreover, these concerns bear immediate attention, even if wide-scale commercialization in some sectors is still years off.

In this volume, we offer preliminary, though certainly not premature, thoughts centered around two central questions: First, how does government confront conditions of acute uncertainty about environmental and health risks? Second, given such uncertainty, how does government structure its relationship with the regulated? To shed light on these questions, this book brings together an array of scholars to ponder lessons from experience in dealing with environmental consequences of technological development, ranging from post–World War II era chemical pesticides to late-twentieth-century genetically modified organisms. These scholars address a set of key issues, including the dilemma of regulating under conditions of uncertainty, the possibilities and limitations of business self-regulation, the organizational capacity of the EPA to adapt to challenges posed by new technologies, and the strengths and weaknesses of regulatory federalism in the United States.

Our focus in this volume is on those areas of environmental and human health most likely to fall within the jurisdiction of federal and state environmental and public health agencies. This distinction is in many ways an artificial one, given that many of these same challenges to government capacity confront other frontline regulatory agencies; for example, nano-medical therapeutics and devices fall within the primary jurisdiction of the U.S. Food and Drug Administration, and nano-enhanced consumer products within that of the Consumer Product Safety Commission. These agencies face an array of challenges compelling enough in their own right to warrant separate studies, even as many of the lessons set forth in this volume apply more broadly. We also confine our immediate attention to the United States, although we acknowledge that any emerging U.S. nanotechnology regulatory regime will be informed by similar discussions taking place in the European Union and elsewhere around the globe, and not only in advanced industrial societies, where the bulk of current research and development takes place. In short, this volume is but a first take on a more expansive agenda of inquiry.

OUTLINE AND CHAPTER SUMMARIES

Contexts

We begin by setting the broader contexts within which debates over the proper form of environmental regulatory approaches to nanotechnology are taking place. In Chapter 2, Sean O'Donnell and Jacqueline Isaacs examine nanotechnology first as an area of exciting scientific inquiry and technical manipulation of matter; then as an inchoate yet highly promising (maybe even "revolutionary") sector with multiform economic, environmental, human health and longevity, and other material benefits; and finally, as a set of emerging concerns about proximate and long-term health and environmental effects of nanomaterials. Their chapter is intended to give nonexperts— which we assume to mean most readers—a basic understanding of the science involved, current and potential applications, and reasons why this study is appropriate.

Central Questions

If policymakers did want to act proactively to address the challenges posed in Chapter 2, does government have the capacity to do so? To address this core question, the book moves on to practical considerations obtained from experience. In Chapter 3, Marc Eisner considers the role of uncertainty in environmental policy in light of the possible challenges posed by nanotechnology. Responses to uncertainty can take multiple forms. One is procedural, as when policymakers try to manage uncertainty through institutional and regulatory design decisions that either empower regulators to act or constrain bureaucratic discretion in ways that limit their capacity to adapt to changing regulatory challenges. Other responses focus on what to regulate. Unable to manage the universe of environmental and health risks, policymakers set priorities, or at least focus limited resources on some subset of risks. Controversy then ensues over the decision rules or metrics applied to make these tough decisions. Regulators also must decide how much to regulate, what constitutes a reasonable level of risk, and how such determinations should be made. Finally, regulators must make decisions about how to regulate, including which regulatory instruments are appropriate given the features of the challenges they face. Yet under conditions of extreme uncertainty, it may be impossible to identify and prioritize problems, design appropriate regulatory responses, and evaluate performance. Thus an ongoing concern is how regulators manage uncertainty under existing legal and institutional constraints, with consequent ramifications for the regulation of nanotechnology and its effects.

Chapter 4 follows logically, in which Cary Coglianese ponders how the absence of prescriptive information about the health and environmental effects of nanotechnology challenges the relationship between regulators and the regulated. In this one respect, nanotechnology is not so novel, Coglianese argues, as regulators in almost all contexts find themselves needing more information than they have and consistently are at a comparative informational disadvantage vis-à-vis the regulated industry. Extrapolating from several streams of research literature, Coglianese suggests that the current lack of information about the full environmental and health effects of nanotechnology creates a strong case for engaging with and delegating to industry—even if that same informational asymmetry makes it difficult to ensure that business will act in a socially optimal manner. In the end, the forms of engagement or delegation that may appear to be attractive in light of the underlying information deficit will likely face inherent difficulty in ensuring effectiveness in regulating nanotechnology—or at least it will be difficult for regulators or the public to know whether these approaches are effective.

Institutions

Having more and better information presumes some institutional capacity to use and apply it. The book next looks at two institutional venues that will have significant impacts on nanotechnology development and regulation. Marc Landy, in Chapter 5, assesses whether EPA as an institution is positioned to handle the coming generation of nanomaterials. Landy first asks what a fully capacitated EPA might look like, and then judges the agency's capabilities in light of that standard. Capacity has many dimensions, even assuming adequate funding. To pursue a multifaceted regulatory mandate, an agency must have the right mix of expertise, grants of authority, institutional design, and leadership characteristics. Regulators also must be able to ask the right questions if the agency is to generate meaningful, transparent, and cogent regulatory answers. Landy expresses doubt that EPA is able to do any of these things, for reasons related to the path-dependent legacy of its formation 40 years ago. He concludes, however, that the acknowledged promises of nanotechnology pose a unique opportunity for a "grand bargain" between industry officials and environmental protection advocates on a balanced conception of risk, one based on the "weight of evidence" and not on a priori presumptions of any particular substance's innocence or guilt. Such a political bargain among long-entrenched and mutually wary policy stakeholders may enable EPA to do its job better, if only the political will to broker such a bargain exists.

And what about state governments? As we sometimes forget, the United States is a federal system, and state governments wield considerable influence in the formation and implementation of a range of federal and state environmental and health policies. Barry Rabe, in Chapter 6, considers nanotechnology as a challenge for all levels of government in a federal system, with a particular focus on the evolving role of states. Rabe begins with an overview of developments in American intergovernmental relations, including factors that today give state governments a significant role in key areas of environmental policy formation and implementation. He then addresses state incentives to get involved in nanotechnology, reflecting both strategic decisions concerning economic benefits from nanotechnology development and possible environmental health risks to residents. In doing so, Rabe devotes particular attention to California, Massachusetts, Minnesota, and Wisconsin, states that have already expressed some level of commitment to understanding and possibly mitigating any potential harm posed by nanomaterials and their processing. He concludes with a general discussion of the challenges and opportunities presented by placing nanotechnology into an intergovernmental context.

Conclusions

Finally, in Chapter 7, Christopher Bosso and W. D. Kay synthesize the findings offered in the preceding chapters and reflect on how government might approach the environmental challenges posed by nanotechnology and other emerging technologies. Does contemporary government have the capacity to handle these challenges? Whatever their broader views about government, citizens expect it to provide some protection from risks they cannot personally control, and to be responsive to public concerns in doing so. Failure can undermine public trust in democratic government. The lesson for nanotechnology from past technological revolutions is that governments must be anticipatory and resilient if they are to protect the public interest. They must be able to project ahead and adapt quickly to the unforeseen. What those goals mean in practice, and current prospects for meeting them, remain contended issues.

REFERENCES

Bosso, Christopher. 1987. *Pesticides and Politics: The Life Cycle of a Public Issue.* Pittsburgh, PA: University of Pittsburgh Press.

Bosso, Christopher, Jackie Isaacs, William D. Kay, Ronald Sandler, and Ahmed Busnaina. 2006. *Nanotechnology in the Public Interest: Regulatory Challenges, Capacity, and Policy Recommendations.* National Science Foundation SES #0609078, 2006–10.

CEIN (Center for Environmental Implications of Nanotechnology). 2006. Program Solicitation. www.nsf.gov/pubs/2007/nsf07590/nsf07590.htm (accessed February 22, 2009).

Daughton, Christian G., and Thomas A. Ternes. 1999. Pharmaceuticals and Personal Care Products in the Environment: Agents of Subtle Change? *Environmental Health Perspectives* 107 (December): 907–38.

EOP (Executive Office of the President), Office of Science and Technology Policy, National Science and Technology Council, Committee on Technology, Interagency Working Group on Nanoscience, Engineering, and Technology. 2003. *The National Nanotechnology Initiative: Leading to the Next Industrial Revolution.* Washington, DC: OSTP.

NNI (National Nanotechnology Initiative). 2008. Funding. www.nano.gov/html/about/funding.html (accessed March 19, 2009).

Roco, Mihail. 2003. Nanoscale Science and Engineering at NSF. National Science Foundation grantees workshop. December. www.nseresearch.org/2003/NewFiles/Over00_NNI_Roco_NSE2003.pdf (accessed March 19, 2009).

SME (Society of Manufacturing Engineers). 2006. Conference: The Next Industrial Revolution: Nanotechnology & Manufacturing. www.sme.org/cgi-bin/get-event.pl?--001640-000007-020103--SME- (accessed March 19, 2009).

Tarrow, Sydney. 2000. Mad Cows and Social Activists: Contentious Politics in the Trilateral Democracies. In *Disaffected Democracies: What's Troubling the Trilateral Countries?* edited by Susan J. Pharr and Robert D. Putnam. Princeton, NJ: Princeton University Press, 270–90.

U.S. Congress. 2003. *21st Century Nanotechnology Research and Development Act*. Public Law 108-153, 108th Cong.

Wardak, Ahson. 2003. *Nanotechnology & Regulation: A Case Study Using the Toxic Substances Control Act*. Washington, DC: Foresight and Governance Project, Woodrow Wilson International Center for Scholars.

White House. 2000. *National Nanotechnology Initiative: Leading to the Next Industrial Revolution*. Washington, DC: White House, Office of the Press Secretary. clinton4.nara.gov/WH/New/html/20000121_4.html (accessed March 19, 2009).

CHAPTER 2

A WORLD OF ITS OWN?
Nanotechnology's Promise—
and Challenges

Sean T. O'Donnell and Jacqueline A. Isaacs

The story of nanotechnology is about the discovery of one world and the possible remaking of another. It is about a technological platform for scientific discovery and innovation. It is about applications with the potential to vastly improve our daily lives. The potential of nanotechnology stirs the imagination, whether through the promise of breakthrough treatments of illness and disease; new modes of energy production, transmission, and storage, with parallel reductions in natural resource use; a cleaner environment, through the application of novel sensing and remediation technologies; or the transformation of agricultural practices to lay the foundation for a twenty-first-century Green Revolution.

To pursue these dreams, scientists and engineers have embarked on an expedition into an essentially invisible and seemingly impossible realm—a tiny world not much bigger than an atom, yet much smaller than a living cell. Working at the nanoscale, scientists and engineers look to manipulate matter from approximately 1 to 100 nanometers (1 to 100 billionths of a meter). In this realm, between the subatomic particles guided by the tenets of quantum mechanics and the larger macro world guided by the more familiar rules of Newtonian physics, emerges a quantum-mechanical regime in which the dynamics of chemical and physical laws have yet to be fully understood. At the nanoscale, new properties emerge at the level of physics, chemistry, and biology, and it is the discovery and development of such new properties that excite scientists and engineers alike. Indeed, bulk chemicals when reduced to their nano size do not always behave in the same manner, or even look the same, due in part to changes in the surface area per unit of volume at the nanoscale. Zinc oxide, for example, is usually opaque in bulk and has long been used to whiten paints and in sunscreens to block out harmful ultraviolet (UV) rays. Nanoparticles of zinc oxide, however, are transparent because of light-scattering effects, and new, nearly transparent sunscreens containing zinc oxide nanoparticles are already on the consumer market. Similarly, the familiar color of gold can show up in different hues of red, blue, or green as a result of slight changes in the size of the nanocluster.

Other properties can change as well. Materials can become more reactive at the nanoscale because of increased surface-area-to-volume and surface-area-to-mass ratios. For example, bulk platinum is a catalyst used frequently in automobile exhaust systems to convert hydrocarbons and carbon monoxide into more harmless emissions, thereby reducing air pollution. When reduced to the nanoscale, with significantly more surface-area sites for chemical reaction, much less of this expensive precious metal is required to achieve the same results. Aluminum is stable in bulk form but becomes explosive at the nanoscale, again as a result of increased surface-area effects. Such nanoaluminum particles may prove particularly useful in producing more powerful, efficient, and even safer rocket propulsion.

The interplay of atomic forces at the nanoscale also changes, yielding new structures for common materials such as carbon, resulting in dramatic improvements in material properties. Until 1991, carbon was known to exist in two structures: diamond cubic and graphite. With the discovery of the fullerene structure, C_{60}, essentially a molecule with 60 carbon atoms bonded in a soccer ball–like shape, ideas for new applications using the resulting enhanced properties abounded. Further exploration of processing techniques resulted in the discovery of carbon nanotubes (CNTs), with shapes analogous to hollow tubes of chicken wire but with diameters on the order of a few nanometers. The thermal conductivity, electrical resistance, and tensile strength of CNTs show dramatic improvements over those of other carbon structures. Carbon nanotubes, for example, are better conductors than metals, because the flow of electrons follows quantum expectations, behaving more like a wave than a particle. Laws of nature that are seemingly irrelevant at the macro level, such as van der Waals forces governing the attraction and repulsion between molecules, are found to feature prominently at the nanoscale, heralding entirely new ways to manipulate matter. Research on van der Waals effects at the nanoscale, for example, has shed light into the ability of the gecko, a small tropical lizard, to cling to smooth surfaces and is paving the way for nanoadhesives of incredible strength (Banegas 2008).

Although theoretical understanding of the nanoscale has been around for decades, the invention of the scanning tunneling microscope (STM) and atomic force microscope (AFM) in the 1980s finally made it possible for scientists and engineers to explore the characteristics of nanomaterials and devices. These tools permit researchers to visualize and, more important, manipulate structures at this level. Similarly, scientists and engineers have been able to fashion nanostructures such as fullerenes (C_{60} also known as buckminsterfullerenes, or buckyballs, after famed architect and futurist Richard Buckminster Fuller) and dendrimers (large, complex molecules with repeatedly branching shapes) as potential delivery systems and scaffolding components at the nanoscale. At this scale, proteins are copied and assembled, enzymes catalyze reactions, energy is transferred in and out of cells, bonds are made and broken, and information such as chemical and electrical activity is stored, released, and transmitted. To put the nanoscale in perspective, these manipulations are happening on the same scale as the more familiar deoxyribonucleic acid (DNA), the basic building block of life, which has a diameter of only a few nanometers. At this level, basic elements and organic compounds begin to organize into the more complex structures necessary for the support of life. So it is not too much to say that nanotechnology has found, in the memorable words of famed physicist Richard Feynman (1960), "plenty of room at the bottom."

Not surprisingly, then, the rich potential of applications at the nanoscale has captured wide attention and enthusiasm. Researchers around the world have begun to recognize that nanotechnology pushes inquiry across the conventional borders of scientific disciplines, and their governments are fostering multilateral initiatives to pool resources and knowledge in the advancement of nanotechnology. In the United States, the federal government's National Nanotechnology Initiative promotes and funds multidisciplinary collaborations among scientists and engineers from diverse fields across academia, industry, and government (NNI 2009). Propelling this initiative is the very real belief that advances in nanotechnology will reap huge technological and economic advantages in energy, agriculture, medicine, structural materials, sensors, computing, pollution remediation, and military applications, including defense against terrorism. Indeed, fear of not keeping apace of developments in nanotechnology has generated a new sort of "arms race" among nations—and, as Rabe notes in Chapter 6, among U.S. states—for economic and technological superiority.

For environmental scientists and engineers, the promises of nanotechnology may be realized through novel pollution prevention and remediation approaches. Many believe that the unique properties of nanomaterials can be harnessed to create highly accurate and predictive biosensors and produce new advances in the treatment and remediation of pollutants, using technology to address long-standing environmental problems (Baeumner 2003). Beyond remediation, they envision the development of green nanotech manufacturing and engineering processes leading to more ecologically sensitive production and waste-treatment systems, substitutes for toxic materials, more efficient and environmentally sustainable agricultural practices, practical biofuels, and enhanced and real-time environmental monitoring.

Thus nanotechnology holds much promise at many levels, yet questions persist as to whether the release of such small particulates into the environment can be tolerated by ecological and human systems. With more than 1,000 consumer products using nanomaterials already on the market (PEN 2009) and many more under development, concern is growing that our efforts to understand the environmental and human health effects of nanomaterials throughout their life cycles may be lagging, especially as potential chances of exposure increase.

Hailed by many as the "next Industrial Revolution," nanotechnology promises social and economic changes writ large. Investors are looking for payoffs in new markets of nano-driven industries and are expecting significant, if not disruptive, changes in supply chains. Given its huge scope of applications, nanotechnology may be transformative, with implications for meeting many of our current global technological challenges.

THE NEXT GREEN REVOLUTION?

The "next Industrial Revolution" may also produce the "next Green Revolution," as scientists, engineers, and industries embrace the challenge of developing sustainable energy, agricultural, and environmental products and practices. In this regard, nanotechnology may offer options for development of sustainable energy sources, less ecologically destructive agricultural practices, and a cleaner overall environment.

Sustainable Energy

Nanotechnology currently has four principal applications in energy research: solar power, fuel cells, biofuels, and batteries. Although solar power makes use of a seemingly endless and free source of energy, its mass production is hampered in large part because of its high manufacturing and production costs. It currently works well where small, contained amounts of energy are needed, such as in calculators, sailboats, or satellites. Nanotechnology as applied in the field of photovoltaics holds the promise of thinner, more easily produced cells that have greater efficiency with use of ambient light. For example, recently developed nanocups, which are a metamaterial (a material that gets its specific properties not from its chemical composition, but from its structure), are capable of capturing light from all directions and refocusing it into one place, thereby eliminating the costly machinery of current thermal solar power used to reposition solar cells to track changing directions of natural light (AZoCleantech 2009). Other developments in photovoltaics include solar cells that can be woven into fabrics and worn, or integrated directly into home building materials, such as glass windows and solar roofing. Such personal use of nano-powered products would reduce the load on the current energy grid and may even aid in producing more efficient energy networks.

Another promising energy application is in the development of hydrogen and methanol fuel cells. Scientists and engineers have begun to make use of the increased surface area of nanomaterials to generate nanocatalysts that produce hydrogen ions from hydrogen and methanol fuels (Mathiesen 2006). While nanoscale platinum is currently used and is more efficient than bulk platinum catalysts, scientists and engineers continue to seek lower-cost, more sustainable alternatives, resulting in the development of more efficient hydrocarbon membranes for fuel cells (Quantum Sphere 2009). These membranes are located where the principal electrochemical reaction occurs, generating electricity directly from the reaction with the fuel (PolyFuel 2009). As they get thinner, these membranes become more energy efficient, because by-products can transfer back across the barrier to be used in the reaction again. This ease of flow decreases the need for costly pumps to perform the process.

Biofuel research and development are similarly looking to nanotechnology and catalysis to create new sources for fuel. The current focus is on efficient production of biofuels from biomass, and the U.S. Department of Energy has teamed up with researchers to exploit nanoparticles to extract oils from algae for use as biofuels (Ames Laboratory 2009). Ideally, nonfood sources such as algae could be used for fuel, reducing the time and space needed for growth and cultivation of traditional biofuel sources such as corn or soy, a practice under growing criticism for its heavy reliance on chemical fertilizers and pesticides, which have deleterious long-term impacts on soil and groundwater. The search for renewable energy in biomass will not necessarily be environmentally benign, however, as burning hydrocarbon fuels derived from "green" sources in the end is still equivalent to burning fuels processed from carbon-based sources (Berger 2008).

Nanotechnology may also be useful in producing more efficient batteries that can deliver greater power on demand and recharge in just a fraction of the time it takes current lithium-ion batteries. Coating the surface of an electrode with nanoparticles increases its surface area, which in turn increases the amount of electric current that flows between the compartments of the battery. As such, nanotechnology may offer solutions to the familiar battery problems: the long time for recharging, short life cycle, safety concerns (less flammable electrode material), and poor performance in high and low temperatures. Such applications of nanotechnology may affect not only conventional battery-operated items, but also larger energy systems such as hybrid electric engines and automobiles (Altair Nano 2009).

Detection and Remediation

A radical transformation in energy sources obtained through nanotechnology may not be immediately forthcoming, given the need for extensive and costly basic research. More attainable in the near term are technologies with beneficial impacts on the environment and human health (EPA 2004). For example, use of highly sensitive nano-enabled biosensors can take laboratory diagnostics out into the field and improve detection and treatment of environmental toxins. Such "lab on a chip" applications will make possible real-time assessment and remediation of a wide array of compounds at extraordinarily low concentrations under difficult and extreme field conditions, including ultratrace levels of heavy metal ions and other radioactive compounds, chemical pesticides, and even antibiotics in milk. Similar sensors are entering the medical field, offering enhanced diagnostics for breast cancer (Science Daily 2008), and the Food and Drug Administration is reviewing an application for a device that would detect very low levels of the anthrax toxin (FDA 2009; Tang et al. 2009).

Investigators also see real promise in dendrimers—artificial, polymer-based molecules with a high level of surface area for absorption because of their unique tree-like shapes, which shoot off in multiple directions from a central core—to isolate and capture specific toxins that bind to them. These nanoparticles may be of particular use in marine environments, trapping various toxins, such as algal toxins, that threaten shellfish populations (Cardona et al. 2002). Other nanoscale detection devices will allow researchers to measure the concentration present in microbial pathogens by using micron-scale piezoelectric cantilever arms that can "weigh" the quantity of molecules from only a small sample (Goldman and Coussens 2005).

Nanotechnology also promises more affordable and efficient approaches to remediation of air and water pollution. Use of nanostructure membranes that can capture exhaust upon emission makes it possible to alleviate air pollution without the need for expensive retrofitting (Engineer Live 2009; Green Technology Forum 2008.). Many new catalysts in the form of nano-crystals are capable of denaturing carbon dioxide (CO_2) and methane (CH_4) gases, filtering them before they are released into the environment. Similarly, researchers are investigating the use of such catalysts to treat existing problems of groundwater contamination in order to stabilize pollutants, such as organic solvents, pesticides, fertilizers, and heavy metals, and remediate some of the most pervasive and problematic groundwater pollutants: chlorinated organics. In this regard, nanotechnology may enable the production of lower-cost-base material, instrumental in remediation efforts. For example, the U.S. Environmental Protection Agency (EPA) has reported promising results with the use of iron nanoparticles to capture chlorinated organics found in contaminated groundwater sites (NCER 2007).

Most dramatic, perhaps, are proposed uses of nanoscale applications to geoengineer the planet and thereby reverse the effects of global warming, whether through wide-scale deployment of nanoscale aerosols in the atmosphere to deflect sunlight, mechanically capturing carbon dioxide via the construction of forests of artificial "trees," or absorbing it "organically" by spreading nanoiron particles into northern ocean areas to stimulate plankton growth (Begley 2007). These ideas are controversial, yet the seriousness with which they are being discussed, even in official circles, indicates the degree to which nanotechnology has enabled dreams once relegated to popular science magazines (Homer-Dixon 2009).

UNCERTAIN EFFECTS

Despite all the promise that nanotechnology holds, any new technology, particularly one operating at the molecular level, also presents possible risks. Indeed, many of the new and exciting properties that make nanotechnology so potentially productive and powerful also make the long-term implications of this technology unpredictable and uncertain. For example, the higher level of reactivity and increased catalytic capacity of nanomaterials that excite scientists and engineers may also pose potentially serious environmental, health, and safety (EHS) problems on release of nanomaterials into the environment. Given their affinity for interaction, small size, and increased mobility, it may be possible for nanomaterials to interact with ecosystems at different entry points—through water, soil, waste streams, or contact with microbial organisms, possibly disrupting the bottom levels of aquatic and animal food chains (AZoNanotechnology 2008; Cimitile 2009).

While mass-based doses are important for nanomaterials, this characterization alone might be problematic for understanding their potential toxicity and designating them as analogous to existing chemicals (see Chapter 3). Other properties also have been found to influence resulting toxicity, including chemical composition, particle size and distribution, concentration, shape, surface area, surface-to-volume ratio, surface chemistry, surface contamination, surface charge, crystal structure, porosity, solubility, state of dispersion and agglomeration, adhesion, production process, and heterogeneity.

With so many potential entry points into the ecosystem and numerous properties to investigate for toxicological response, the extent of risk assessment needed is daunting. Realistically, no decision for use or release of a nanomaterial into the environment, even with substantial testing, will be made without some risk and degree of uncertainty. Indeed, this condition of uncertainty is made all the more acute by the fact that initial releases of nanomaterials will likely be at very low concentrations, so it could take years before any cumulative effects demonstrate a material's toxicity. Similarly, it is then likely that the true affinities and effects of any material will become apparent only post hoc. That is, we may not know the best tests to run until after the material has been released and its particular pathways become evident.

Moreover, the vastly decreased particle size of nanomaterials increases the problem of uncertainty. Currently, our sense of ecological and physiological pathways is a result of how we understand the interactions of micro-size and larger materials. The barriers and boundaries that exist at the micro level may not be as effective at the nanoscale and not good predictors of how nanoparticles will migrate. In fact, settled ideas of transport and fate may well need to be reconsidered, as nanoparticles may find different routes for transfer and easily could go farther than other types of materials. Even life-cycle notions about materials at end of use might need to be readdressed as we learn more about what it takes to make a nanoparticle relinquish its reactive properties and what conditions may reactivate them (Zhu et al. 2007).

It is also clear from the variety of scientific journals dedicating special issues to nanotechnology that the topics of sustainability and environmental health and safety of emerging nanotechnologies are beginning to rouse significant interest in professional communities (e.g., Clift and Lloyd 2008; Helland and Kastenholz 2008; Randles et al. 2008; Tinkle 2008). Such concerns also have gradually taken their place on government research agendas. On September 21, 2006, the U.S. House Committee on Science expressed its desire for greater federal research on and coordinated action to address the potential EHS risks of nanotechnology (U.S. Congress 2006). That same day, not coincidentally, the National Science and Technology Council (NSTC) of the U.S. National Nanotechnology Initiative released its EHS strategy on the research and information needs deemed essential to ensure the responsible development of nanotechnology in the decade to come (NSTC 2006). These reports echoed a U.K. government-commissioned report (RS & RAE 2004) and other public calls for increased attention to environmental health and safety issues related to emerging nanotechnologies (Maynard 2006; Maynard et al. 2006). A subsequent strategy for prioritizing federal EHS research was compiled by the NSTC's Nanotechnology Environmental and Health Implications Working Group (NEHI 2007), while EPA concurrently developed a roadmap for its own EHS research needs (EPA 2007). Based on these documents, in 2008 EPA's Office of Research and Development program in nanomaterial research identified four key themes to provide leadership for federal research on EHS issues: sources of, fate, transport, and exposure to nanoparticles; human health and ecological research to inform risk assessment and test methods; assessment methods and case studies; and preventing and mitigating risks (EPA 2008). Although now considered high priority with the establishment in 2008 of two EPA–NSF Centers for Environmental Implications of Nanotechnology, such EHS research will take years to yield definitive results, even assuming adequate funding in the United States as well as similar efforts in other countries.

This being said, we already can make deductions about more proximate impacts. Even amid general uncertainty, experience with prior technologies suggests that those most likely to first experience any ill effects will be those exposed during product fabrication, rather than consumers who purchase or use the products (Bass 2008). One particularly sobering episode came to light in an August 2009 clinical study distributed globally by the Reuters news service (Lyn 2009), concerning the deaths of seven Chinese paint-factory workers who had worked for months with nanoparticles. Although its author acknowledged that the workers had operated in a confined space and lacked proper protection, such as respirators, the clinical study stirred considerable discussion in professional circles because it asserted possible detrimental health effects of nanoparticles to humans.

Not surprisingly, policymakers and industry leaders have focused con-siderable efforts on establishing guidelines for safe practice in working with nanomaterials. At the federal level, the National Institute for Occupational Safety and Health has published summaries of findings to date, highlighting in particular concerns about the abilities of these materials to penetrate dermal barriers, cross cell membranes, travel neuronal pathways, breach the gas-exchange regions of the lung, travel from the lung throughout the body, and interact at the molecular level (NIOSH 2007; NIOSH 2008).

Similar efforts are taking place at an international level, particularly through nongovernmental organizations. In 2006, the Organisation for Economic Co-operation and Development (OECD) formed a Working Party on Manufactured Nanomaterials to foster international cooperation on health- and environmental safety-related aspects of manufactured nano-materials (OECD 2009). One new organization, the International Alliance for NanoEHS Harmonization (IANH), seeks to foster collaboration among scientific experts worldwide to establish protocols for reproducible toxico-logical testing of nanomaterials in both cultured cells and animals (IANH 2009). This effort is motivated in part by studies indicating that key gaps exist in scientific knowledge regarding the biological interactions with nano-particles and subsequent toxicological responses. Two significant limitations to resolving these issues are the lack of testing protocols that enable repro-ducible assessment of the biological interactions of nanoparticles with cells and animals; and the lack of correlations between interactions observed in cells and in animals. In an effort to establish universal standards for nano-materials research and risk assessment, and enable researchers to compare and contrast work conducted at laboratories around the world, formation of the IANH was encouraged by a number of government agencies, including the U.S. National Science Foundation, National Institutes of Health, National Institute for Occupational Safety and Health, and National Institute of Standards and Technology; the Seventh Framework Programme of the European Community for research, technological development, and demonstration activities; and the Japanese National Institute for Materials Science (Nanowerk News 2008).

THE CHALLENGE

Not all risk or uncertainty about nanotechnology stems from a lack of scientific knowledge or an inability to measure and assess the environmental, health, and safety impacts of nanomaterials. As ethicist Ronald Sandler (2009) reminds us, social, political, and economic factors significantly shape the possibility of nanotechnology developing in a safe and "responsible" manner. In underscoring the social dimension of uncertainty and risk, we highlight the extent to which control over nanotechnology is not merely a scientific problem, but one infused by economic, political, and ethical considerations. In this way, nanotechnology emerges within specific contexts that may require decisionmakers to consider the political nature of uncertainty itself. Indeed, given the rate at which industry applications now outstrip toxicological studies, other factors beyond scientific results may need to be considered. Simply put, as Marc Eisner and Cary Coglianese both address (see Chapters 3 and 4), any decision to regulate nanotechnology may need to come at a time well before the full evidence is in.

As such, the rush to launch nanotechnology as not merely a technological force, but also an economic one, increases its potential environmental and health risks. In a time of economic downturn, this becomes especially problematic, because innovation is spotlighted as an urgent matter of economic solvency. After World War II, for example, asbestos was confidently seen as an all-purpose material for rebuilding war-torn countries, and only decades later did its dangers become widely recognized (Tweedale 2001). Notably, nanotechnology has been consistently identified with securing American economic advantage on the global scale, and its economic potential has been repeatedly stressed as a viable economic engine in not just one, but many different markets. While such fervor concentrates attention and resources, it also creates the conditions for heightened risk-taking and the possibility of unforeseen mistakes.

Nanotechnology has important opportunities to meet technological challenges in such diverse areas as electronics, energy, water purification, food storage, and therapeutics. These emerging technologies hold great promise for both global economic growth and a sustainable environment for human welfare. But because the effects of nanomaterials' interactions with the environment and living systems are largely unknown and difficult to predict, nanotechnology's potential risks are all the more difficult to ascertain. Moreover, the speed with which the nanotechnology market is developing threatens to outstrip the capacity of science to understand its myriad risk profiles. The market sales for intermediate products with nano-scale features—coatings, fabrics, memory and logic chips—alone is estimated to grow from approximately $70 billion in 2005 to $850 billion by 2014 (Berger 2007). At the higher end of the value chain, applications spanning industrial, consumer, and medical products exceeded sales of $147 billion in 2007 and are predicted to give rise to $3.1 trillion in manufactured goods by 2015 (Lux Research 2008). These estimates suggest an increasing possibility for potential human exposure and environmental dispersion before we can gain a firm understanding of the risks involved.

The above data are only for nanoproducts about which we know something. Not all industries or governments are forthcoming about all uses of nanomaterials. "Stealth" nanoproducts may exist for a number of reasons, including the desire to maintain manufacturing secrecy, market advantage, or national security (Carafano and Gudgel 2007). Some manufacturers contend that embedding nanomaterials in larger components renders them harmless, yet there is no guarantee how long or to what degree a nanoparticle or nanostructure will stay embedded, particularly once it enters the waste stream. As a result, many products will be released into the market—and the environment—without undergoing any scientific or regulatory scrutiny until means of detecting nanoparticles are developed.

Along with the "legitimate" release of any currently undetectable nano-particles by industry, the potential also exists for their targeted release as a tool for bioterror or ecoterrorism. To date, it has been difficult for rogue states and regimes to have the same level of sophisticated weaponry as developed countries, such as nuclear weapons, because many of the resources are scarce and costly. The relatively greater accessibility to and low cost of nanomaterials, however, make their use in bioterror or ecoterrorism more likely. As nanotechnology globalizes, so too does the potential for the development of untraceable weapons of mass destruction. Such scenarios only increase environmental risk and uncertainty.

Clearly, nanotechnology's potential impacts on the environment and human health are not merely technical matters or simple research problems. Both the excitement and uncertainty surrounding nanotechnology are grounded in social, political, and economic realities that must become part of our calculus about accepting and managing risks. Decisions regarding acceptable risk must be made not only while we are still defining the problems and challenges, but also while we are articulating the sort of world we wish to live in. Under these conditions, addressing environment, health, and safety concerns is an ethical challenge in which the uncertain realm of nanotechnology becomes our own.

REFERENCES

Ames Laboratory. 2009. Nanofarming Technology Harvest Biofuel Oils without Harming Algae. www.nanotech-now.com/news.cgi?story_id=32791 (accessed April 20, 2009).

Altair Nano. 2009. What Can Nanotechnology Do for Your Batteries? www.b2i.us/profiles/investor/fullpage.asp?f=1&BzID=546&to=cp&Nav=0&LangID=1&s=236&ID=9307 (accessed April 20, 2009).

AZoCleantech. 2009. Metamaterial Made Using Nanotechnology Could Lead to Highly Efficient Solar Cells. www.azocleantech.com/Details.asp?newsID=4919 (accessed April 19, 2009).

AZoNanotechnology. 2008. Nanoparticles and the Health of the Soil Ecosystem. www.azonano.com/news.asp?newsid=8193 (accessed April 20, 2009).

Baeumner, Antje J. 2003. Biosensors for Pollutants and Food Contaminants. *Analytical and Bioanalytical Chemistry* 377 (3): 434–45.

Baker, Stephen, and Adam Aston. 2005. The Business of Nanotech. *Business Week*, February 14. www.businessweek.com/magazine/content/05_07/b3920001mz001.htm (accessed August 25, 2009).

Banegas, Diane. 2008. How to Make Adhesive as Good as a Gecko. www.nsf.gov/discoveries/disc_summ.jsp?cntn_id=112442&org=NSF (accessed August 29, 2009).

Bass, Carole. 2008. As Nanotech's Promise Grows, Will Puny Particles Present Big Health Problems? *Scientific American*, February 15.

Begley, Sharon. 2007. The Geo-engineering Scenario: Why Even a Desperate Measure Is Starting to Look Reasonable. www.newsweek.com/id/71691 (accessed August 29, 2009).

Berger, Michael. 2007. Debunking the Trillion Dollar Nanotechnology Market Size Hype. *Nanowerk*, April 18. www.nanowerk.com/spotlight/spotid=1792.php (accessed November 30, 2009).

Carafano, James, and Andrew Gudgel. 2007. Nanotechnology and National Security: Small Changes, Big Impact. www.heritage.org/Research/NationalSecurity/bg2071. cfm (accessed April 20, 2009).

Cardona, Claudia M., Stephan H. Jannach, Hao Huang, Yukiko Itojima, Roger M. Leblanc, Robert E. Gawley, Gary A. Baker, and Eric B. Brauns. 2002. Spatially Resolved Derivatization of Solid-Phase-Synthesis Beads with Fluorescent Dendrimers: Creation of Localized Microdomains. *Helvetica Chimica Acta* 85 (10): 3532–58.

Cimitile, Michael. 2009. Nanoparticles in Sunscreen Damage Microbes. www.sciam. com/article.cfm?id=nanoparticles-in-sunscreen (accessed April 20, 2009).

Clift, Roland, and Shannon Lloyd, eds. 2008. Special Issue on Nanotechnology and Industrial Ecology. *Journal of Industrial Ecology* 13 (September).

Engineer Live. 2009. *New Membranes Design Will Improve Carbon Dioxide Capture.* www.engineerlive.com/Oil-and-Gas-Engineer/Environment_Solution/New_ membranes_design_will_improve_carbon_dioxide_capture/19942/ (accessed April 20, 2009).

EPA (U.S. Environmental Protection Agency). 2004. Nanotechnology Grand Challenge in the Environment: Research Planning Workshop Report Vision for Nanotechnology R&D in the Next Decade. Washington, DC: U.S. EPA.

———. 2007. *Nanotechnology White Paper.* Science Policy Council, Nanotechnology Working Group. es.epa.gov/ncer/nano/publications/whitepaper12022005.pdf (accessed July 16, 2008).

———. 2008. Draft Nanotechnology Research Strategy. Office of Research Development, January 24. www.epa.gov/ncer/nano/publications/nano_strategy_012408. pdf (accessed August 25, 2009).

FDA (U.S. Food and Drug Administration). 2009. FDA Assessing Feasibility of Using Nanotechnology Test to Detect Anthrax Following a Bioterrorist Attack. www. healthnewsdigest.com/news/FDA_Approval_240/FDA_Assessing_Feasibility_ of_Using_Nanotechnology_Test_to_Detect_Anthrax_Following_a_Bioterrorist_ Attack.shtml (accessed April 19, 2009).

Feynman, Richard P. 1960. There's Plenty of Room at the Bottom: An Invitation to Enter a New Field of Physics. *Engineering and Science* 23 (5): 22.

Goldman, Lynn, and Christine Coussens, eds. 2005. *Implications of Nanotechnology for Environmental Health Research.* Roundtable on Environmental Health Sciences, Research, and Medicine. Board on Health Sciences Policy. Institute of Medicine. Washington, DC: National Academies Press.

Green Technology Forum. 2008. New Nano-membrane Captures CO_2. www.greentech forum.net/2008/02/12/new-nano-membrane-captures-co2/ (accessed April 19, 2009).

Helland, Aasgeir, and Hans Kastenholz, eds. 2008. Special Issue on Sustainable Nanotechnology Development. *Journal of Cleaner Production* 16 (May–June).

Homer-Dixon, Thomas. 2009. Uncertainty, Fat Tails, and Time Lags: Why We Must Start Planning Now to Geoengineer Earth Soon. Paper presented at the Annual

Meeting of the American Political Science Association. September 2009, Toronto, Ontario, Canada.

IANH (International Alliance for NanoEHS Harmonization). 2009. Home page. www.nanoehsalliance.org/ (accessed August 25, 2009).

LuxResearchInc.2008.NanomaterialsStateoftheMarketQ32008:StealthSuccess,Broad Impact. July 1. https://portal.luxresearchinc.com/research/document_excerpt/ 3735 (accessed November 30, 2009).

Lyn, Tan Ee. 2009. Deaths, Lung Damage Linked to Nanoparticles in China. www. reuters.com/article/technologyNews/idUSTRE57I1Y720090819?pageNumber=2 &virtualBrandChannel=11611&sp=true (accessed August 30, 2009).

Mathiesen, Ben. 2006. Nano-scale Fuel Cells May Be Closer Than We Think, Thanks to an Inexpensive New Manufacturing Method. www.physorg.com/news11654. html (accessed April 12, 2009).

Maynard, Andrew D. 2006. Nanotechnology: A Research Strategy for Addressing Risk. PEN3, Project on Emerging Nanotechnology, Woodrow Wilson International Center for Scholars, July. www.nanotechproject.org/file_download/files/PEN3_ Risk.pdf (accessed August 24, 2009).

Maynard, Andrew D., Robert J. Aitken, Tilman Butz, Vicki Colvin, Ken Donaldson, Günter Oberdörster, Martin A. Philbert, John Ryan, Anthony Seaton, Vicki Stone, Sally S. Tinkle, Lang Tran, Nigel J. Walker, and David B. Warheit. 2006. Safe Handling of Nanotechnology. Nature 444 (7117): 267–69.

Nanowerk News. 2008. International Group Addresses Lack of Consensus on Nanotoxicology Test Procedures. www.nanowerk.com/news/newsid=7157.php (accessed August 25, 2009).

NCER (National Center for Environmental Research). 2007. Nanotechnology: An EPA Perspective Factsheet. es.epa.gov/ncer/nano/factsheet (accessed April 20, 2009).

NEHI (Nanotechnology Environmental and Health Implications Working Group). 2007. Prioritization of Environmental, Health, and Safety Research Needs for Engineered Nanoscale Materials. www.nano.gov/Prioritization_EHS_Research_ Needs_Engineered_Nanoscale_Materials.pdf (accessed August 25, 2009).

NIOSH (National Institute for Occupational Safety and Health). 2007. Progress toward Safe Nanotechnology in the Workplace. Publication No. 2007-123. Atlanta, GA: Department of Health and Human Services, Centers for Disease Control and Prevention.

———. 2008. Strategic Plan for NIOSH Nanotechnology Research: Filling the Knowledge Gaps. Atlanta, GA: Department of Health and Human Services, Centers for Disease Control and Prevention.

NNI (National Nanotechnology Initiative). 2009. Home page. www.nano.gov (accessed August 24, 2009).

NSTC (National Science and Technology Council). 2006. Environmental, Health, and Safety Research Needs for Engineered Nanoscale Materials. www.nano.gov/ html/news/EHS_research_needs.html (accessed August 24, 2009).

OECD (Organisation for Economic Co-operation and Development). 2009. Working Party on Manufactured Nanomaterials. www.oecd.org/department/0,3355,en_2649_37015404_1_1_1_1_1,00.html (accessed August 25, 2009).

PEN (Project on Emerging Nanotechnologies). 2009. Consumer Products Inventory. www.nanotechproject.org/inventories/consumer (accessed April 20, 2009).

PolyFuel. 2009. PolyFuel Sets New Record for Portable Fuel Cell Performance—Again, New, Ultra-thin 20-Micron Membrane Material Beats PolyFuel's—and Industry's—Best by 40%. www.polyfuel.com/pressroom/press_pr_110706.html (accessed April 20, 2009).

Quantum Sphere. 2009. Next-Generation Fuel Cells. www.qsinano.com/apps_fuelcell.php (accessed April 20, 2009).

Randles, Sally, Paul Dewick, Denis Loveridge and Jan C. Schmidt, eds. 2008. Special Issue on Nano-Artifacts: Converging Technologies at the Nano-Scale: Challenges for Governance, Sustainability, Industry and Research. *Technology Analysis & Strategic Management* 20 (1).

RS & RAE (Royal Society and Royal Academy of Engineering). 2004. Nanoscience and Nanotechnologies: Opportunities and Uncertainties. www.nanotec.org.uk/finalReport.htm (accessed August 24, 2009).

Sandler, Ronald L. 2009. *Nanotechnology: The Social and Ethical Issues. Project on Emerging Nanotechnologies*. Washington, DC: Woodrow Wilson International Center for Scholars.www.nanotechproject.org/publications/archive/pen16/ (accessed August 25, 2009).

Science Daily. 2008. Disease Diagnosis in Just 15 Minutes? Biosensor Technology Uses Antibodies to Detect Biomarkers Much Faster. www.sciencedaily.com/releases/2008/10/081001093231.htm (accessed April 20, 2009).

Tang, Shixing, Mahtab Moayeri, Zhaochun Chen, Harri Harma, Jiangqin Zhao, Haijing Hu, Robert H. Purcell, Stephen H. Leppla, and Indira K. Hewlett. 2009. Detection of Anthrax Toxin by an Ultrasensitive Immunoassay Using Europium Nanoparticles. *Clinical and Vaccine Immunology* 16 (3): 408–13.

Tinkle, Sally. 2008. Special Issue on Nanomaterials in the Environment. *Environmental Journal of Toxicology and Chemistry* 27 (9).

Tweedale, Geoffrey. 2001. *Magic Mineral to Killer Dust: Turner & Newall and the Asbestos Hazard*. Oxford, UK: Oxford University Press.

U.S. Congress. House Committee on Science. 2006. *Research on Environmental and Safety Impacts of Nanotechnology: What Are the Federal Agencies Doing?* Hearing before the Committee on Science, U.S. House of Representatives, 109th Cong., 2nd sess.

Zhu, Ye, Juefei Zhou, and Mark G. Kuzyk. 2007. Two-Photon Fluorescence Measurements of Reversible Photodegration in a Dye-Doped Polymer. *Optics Letters* 32 (8): 958–60. As reported at www.physorg.com/news95408195.html (accessed April 20, 2009).

CHAPTER 3

INSTITUTIONAL EVOLUTION OR INTELLIGENT DESIGN?
Constructing a Regulatory Regime for Nanotechnology

Marc Allen Eisner

Nanotechnology is an emerging technological paradigm (Dosi 1982) with profound implications for fields ranging from material sciences to pharmacology. Comparable to earlier disruptive technologies, such as mass production or integrated circuitry, nanotechnology has the capacity to transform practices and product designs in multiple industries in ways that are difficult to anticipate. The complexity and dynamism inherent in nanotechnology raise important questions about potential environmental impacts and the adequacy of the existing regulatory system. Indeed, at present there are far more questions than answers. Faced with conditions of insufficient data and rapidly evolving scientific theory, regulators may be incapable of identifying underlying causal mechanisms, understanding the health and environmental ramifications of exposure, and designing regulatory responses. Yet, rather than waiting for a regulatory regime to evolve—a process that so often has been shaped by an adverse event or salient crisis—there has been considerable interest within government, environmental and public health advocacy circles, and even segments of the business community in striking a precautionary posture and exploring issues of regulatory design in advance.

"Regulation after an adverse event has its advantages," says Robin Fretwell Wilson (2006), noting that "the event highlights the need for regulation, crystallizes the policy choices we face, allows us to consider real, not perceived costs of regulating or choosing not to regulate, and the regulations we adopt are more meaningful." Considering regulatory design *ex ante*, in contrast, forces one to work without comparable clarity, particularly when exploring a highly dynamic emerging technology. By necessity, therefore, the discussion in this chapter is incomplete and speculative. Reflecting the tenuous nature of the enterprise, the chapter proceeds as a series of questions and reflections on where the answers might lead.

QUESTIONS ABOUT CATEGORIZATION

First, what are we to regulate? At a simple level, the answer is nanotechnology, which in a narrow sense involves the production of materials, structures, or systems between 1 and 100 nanometers (a nanometer is one-billionth of a meter), as per the standard used by the federal government's National Nanotechnology Initiative (NNI n.d.). Materials at this scale are too small to be detected by conventional laboratory microscopes. According to a recent report by the National Materials Advisory Board of the National Academies of Science: "Nanotechnology is not simply about small particles, materials, or products. It is not one type of technology with a defined use. Rather, nanotechnology is an enabling technology that promises to contribute at many frontiers of science and technology" (NMAB 2006). Because nano-technology has a broad range of applications, from material sciences to pharmaceuticals, it is perhaps best understood as an emerging scientific or technological paradigm that will have disruptive and transformative effects in multiple industries, such as manufacturing, energy, communications, health care, consumer products, and defense.

So how do we regulate an enabling technology or, even more abstractly, a technological paradigm? Existing regulations are designed to mitigate the environmental, health, and safety consequences stemming from exposure to or release of specific substances, not paradigms. Should debates over regu-latory design follow suit and focus on the potential implications of exposure to specific substances and structures at the nanoscale, seeking to adjust existing regulatory statutes and instruments to meet the new challenges? Or would such an approach entail the risk of grafting new regulatory responsibilities on to existing policies, thereby failing to develop the capacity to regulate those features of nanotechnology that are genuinely unique?

QUESTIONS ABOUT TIMING

When do we regulate? Is the time ripe for extending current regulations to nanotechnology, or is development at too early a stage for regulators and researchers to understand the nature and magnitude of the regulatory problems? At present, commercial applications are confined to the integration of passive nanostructures, such as metal oxides, quantum dots, or carbon nanotubes, into existing products (batteries, catalysts, cosmetics, food additives, fuel cells, paints) and, to a lesser extent, the development of active nanostructures that change their properties during use—for example, drug-delivery particles that release pharmaceutical agents when they reach their intended destination. But within the next two decades, commercial use is likely to extend to systems of nanostructures, such as nanoscale electronic components that can self-assemble to form circuits, and the development of molecular nanosystems, which have been defined as "heterogeneous networks in which molecules and supramolecular structures serve as distinct devices" (ISO 2007).

The rapid development of nanotechnology is being driven by heavy infusions of public and private financing. In the United States, the National Nanotechnology Initiative has a fiscal year 2010 budget of $1.6 billion spread across 13 government agencies ranging from the Department of Defense to the National Institutes of Health (NNI 2009). State and regional development initiatives provide additional resources and opportunities for collaboration (see Chapter 6). Heavy government funding has been combined with a massive infusion of commercial and private venture capital. By the end of 2007, more than 4,800 patents involving some facet of nanotechnology had been registered with the U.S. Patent and Trademark Office (NSTC 2007). Although the United States has assumed a leadership role in nanotechnology development, global investments have been significant and continue to grow rapidly. In 2005, global government spending was more than $4.5 billion, with an additional $4.5 billion in corporate spending and $497 million in venture capital (ISO 2007).

By contrast to the strategic investment in and attention to technological development and commercialization, the history of environmental regulation reveals a clear pattern of indifference punctuated by a salient event, leading to a regulatory reaction (Baumgartner and Jones 1993; Bosso 1987). New regulatory capacity largely is created after the damage is done, and often in haphazard, incremental, and incomplete spasms of government response. Given this history, it would seem prudent to consider issues of regulatory design at the early stages of technological development and commercialization rather than wait—again—for a crisis to force regulatory change. Yet the potential environmental and health effects raised by the current phase of development of nanotechnology may be qualitatively different than those that will emerge over the course of the next two decades. Thus any reflections on regulatory design will have to be premised on recognition of the dynamics of the underlying technology. Ideally, any regulatory design decisions will have to place a high value on robustness, resilience, and corrigibility if they are to survive in a rapidly changing environment. How to integrate those features into a regulatory regime that also must respond to public demands for democratic accountability and certainty about risk is the obvious challenge.

QUESTIONS ABOUT SCIENTIFIC UNCERTAINTY

What do we know? Nanotechnology has become a subject of great interest, because materials, structures, and devices have novel properties and functions at the nanoscale. The behavior of nanomaterials varies based on size, shape, and surface area (see Chapter 2). Because nanoparticles present a greater surface area relative to volume, for example, they have been of great interest in everything from sunscreens to explosives. To place things in regulatory context, toxicity at the nanoscale may be more a product of surface area than mass. The magnitude of the informational problems becomes clear when we contemplate the various combinations of independent and interdependent variables that affect toxicity, including size, shape, surface characteristics, material composition, and chemical reactivity. To use a simple example, we cannot consider the impact of exposure to one single-walled carbon nanotube, because given the various combinations of relevant physical features, up to 50,000 different permutations are possible, many of which may raise dissimilar environmental and health problems (Davies 2006).

Understanding the health and environmental impacts of exposure to nanoparticles is a daunting task, to be certain. But regulating exposure to substances that may prove toxic involves finding answers to relatively well-defined questions that are not dissimilar to the kinds of questions already engaged by the Environmental Protection Agency (EPA) and Occupational Safety and Health Administration (OSHA) on a daily basis. Unfortunately, these questions, for all their complexity, may be the easiest ones to answer. As the novel properties of nanotechnology are explored in multiple industries, and the research progresses rapidly from passive nano-structures to heterogeneous networks of active nanostructures, the informational problems faced by regulators will increase dramatically—perhaps exponentially—and the policy-based learning that occurs during the earliest phases of technological development may not provide a firm foundation for regulating advanced applications.

QUESTIONS ABOUT RISK ASSESSMENT

How do we judge risk? The risk assessment process is a well-established methodology for identifying hazards, assessing the impact of exposure, and setting regulatory thresholds. It provides the analytical framework within which regulations are developed at EPA and OSHA. Uncertainty is a constant feature of risk assessment. Former EPA Administrator William D. Ruckelshaus (1983, 1026) has accurately described risk assessment as a "shotgun wedding between science and the law." While "science thrives on uncertainty," Ruckelshaus notes, "EPA's laws often assume, indeed demand, a certainty of protection greater than science can provide with the current state of knowledge." Two questions then emerge. First, how does EPA manage this disjunction? Second, are its standard approaches to managing uncertainty sufficient in the area of nanotechnology?

Within the confines of the risk assessment process, regulators must manage problems of model and parameter uncertainty. To be useful, the models employed in risk assessment must generate causal inferences and predictions of the health and environmental consequences of exposure to specific chemical agents. Model uncertainty emerges when gaps exist in the underlying science. EPA's Risk Assessment Task Force notes that model uncertainty "is inherent in risk assessment that seeks to capture the complex processes impacting release, environmental fate and transport, exposure, and exposure response. EPA's models are often incomplete and knowledge of specific processes limited" (EPA 2004a, *33*). Parameter uncertainty, in contrast, is a product of measurement error, sampling error, or missing data. How great are these two forms of uncertainty? According to EPA, its risk estimates are "driven by the uncertainty and variability inherent in practically all the information and methodologies EPA uses" (*141*). The means used to manage these sources of uncertainty can often have a pronounced impact on the final regulatory decisions (Andrews 2006).

EPA adopts various expedients to manage uncertainty and execute its statutory mandates. It manages model uncertainty through the use of default assumptions that reflect professional judgments and the underlying scientific consensus. It manages parameter uncertainty by collecting new data, employing models to generate estimates of missing data, or using surrogate data (EPA 2004a). Although there are clear remedies for various forms of uncertainty, they must be grounded in an understanding of the underlying science. These scientific foundations have yet to emerge in nanotechnology. As EPA's *Nanotechnology White Paper* explains:

> Since we don't have a complete understanding of how nanoparticles behave, each stage in their lifecycle, from extraction to manufacturing to use and then to ultimate disposal, will present separate research challenges. Nanomaterials also present a particular research challenge over their macro forms in that we lack a complete understanding of nanoparticles' scientific properties. (EPA 2005, *62*)

The scientific uncertainty extends to the most elemental questions: "current chemical representation and nomenclature conventions may not be adequate for some nanomaterials" (*33–34*). These problems are not unique to EPA. As Paul Schulte of the National Institute for Occupational Safety and Health (NIOSH) recently noted, "You could probably count the world's published literature on exposure to nanoparticles on both hands" (Bass 2008). The recent creation of two jointly funded EPA–NSF Centers for Environmental Implications of Nanotechnology underscores official recognition of the acute information scarcity and scientific uncertainty.

Setting aside the adequacy of the underlying science, we might ask a far more fundamental question: are the theoretical assumptions underlying the risk assessment process operable at the nanoscale? The risk assessment process presumes a causal relationship between material volume and exposure, on the one hand, and toxicity, on the other. But given the novel properties of nanotechnology, we cannot simply infer toxicity of a nanomaterial from what is known about its macro-scale counterpart. Material volume is no longer the most relevant variable. As Arius Tolstoshev (2006, *21*) points out, "The unique properties and extremely small size of nanomaterials are such that even determining the full extent of the risks to human health and environment is currently beyond the means of existing risk assessment frameworks." Yet "we don't know" will not likely be an acceptable response should public concerns about the safety of nanomaterials be aroused.

QUESTIONS ABOUT THE ADEQUACY OF EXISTING STATUTORY AUTHORITY

What capacity currently exists? EPA's capacity to leverage information and regulate chemicals is structured by existing statutory authority. Is this authority sufficient to develop the information necessary to set regulatory standards and regulate nanotechnology? Marc Landy, in Chapter 5, ponders the broader question of the organizational capacity of EPA. Here, we will look more closely at the agency's regulation of chemicals under the Toxic Substances and Control Act (TSCA) of 1976.

Most chemicals in use are regulated by EPA's Office of Pollution Prevention and Toxics (OPPT) under authority granted by TSCA (Bergeson et al. 2000). The law currently treats existing and new chemicals differently. When the original TSCA Inventory was compiled in 1979, the 61,000 chemicals currently in use were designated as existing chemicals; with periodic updates, the TSCA Inventory currently contains some 83,000 existing chemicals. Under Section 6 of TSCA, EPA is authorized to regulate and even ban the manufacture, processing, distribution, use, or disposal of existing chemicals if it has determined that they pose an "unreasonable risk" to human health or the environment. EPA's regulatory efforts are plagued by the difficulties of accessing the information to substantiate the determination of risk, the efficacy of substitutes, and the economic impacts of the regulatory response. Any one of these requirements poses substantial challenges, and it is not surprising that three decades after TSCA's enactment, the majority of existing chemicals had not undergone basic toxicological testing (EDF 1997; EPA 1998; Denison 2009). To make things more difficult, Congress established the requirement of finding "unreasonable risk" without defining what constitutes an unreasonable risk, thereby creating ongoing opportunities for regulated businesses to challenge the evidentiary foundations of testing rules.

Section 5 of TSCA authorizes EPA to regulate new chemicals, or existing chemicals with "significant" new uses, prior to their manufacture, importation, processing, or distribution. Producers or importers are required to submit a premanufacture notice (PMN) at least 90 days in advance, supported by basic data on the substance, production volume, use, exposure, and whatever information the producer possesses regarding health and environmental effects. The PMN review process functions under enormous constraints (Rhomberg 1996). EPA receives an average of 1,500 PMNs annually, and unless it acts within 90 days, manufacturing and distribution are approved by default. Because firms are not required to submit new substances for testing prior to the submission of a PMN, 67 percent of submissions include no test data, and 85 percent no health data.

During the 90-day review period, the OPPT applies a rough screen based on molecular structure and existing knowledge of structurally related compounds (EPA 2007). Unfortunately, because of the novel behavior of materials at the nanoscale, this knowledge—and thus the current screening process—may not be applicable to nanotechnology. The OPPT can require additional information during the review period, but it must show that the chemical constitutes an unreasonable risk—a determination that is difficult to substantiate in the absence of information.

Complicating matters further, most nanomaterials have the same chemical structure as materials that are already in the chemical inventory. No consensus exists as to whether chemicals with a common chemical identity but different morphology and particle size will qualify as new chemicals under TSCA. In practice, EPA terminates the PMN review process if it determines that the chemical identity is already in the inventory. TSCA Section 5(a) also authorizes EPA to promulgate significant new-use rules, essentially treating chemicals that are currently in the inventory as if they were new. This likely provides EPA with regulatory authority in some, but not all, cases. Yet the promulgation of significant new-use rules could be prohibitively costly if executed on a case-by-case basis. Although TSCA authorizes the issuance of rules for entire categories of chemicals (ABA 2006), the scientific foundations for categorization have yet to be developed. Realistically, given the resource constraints and the facts that most nanomaterials involve the use of inventory chemicals and, regardless of the use, will be used in relatively small amounts, EPA may well determine that the promulgation of significant new-use rules is not a good application of its scarce resources. Indeed, EPA already exercised its discretion to exempt a specific carbon nanotube under the Low Release and Exposure Exemption (Davies 2007).

Outside of the PMN process, TSCA Section 4 authorizes EPA to promulgate rules requiring manufacturers or processors to conduct testing to develop data on health and environmental effects. Such rules must be justified with findings regarding production, exposure, and potentially unreasonable levels of risk that are difficult to substantiate without the very data that the rules would generate. Historically, EPA has used the threat of mandatory testing rules to obtain voluntary testing: "Unless voluntary testing agreements are entered ... EPA would need to demonstrate, through notice and comment rulemaking, that it can support either a risk- or exposure-based finding for a nanoscale substance that is subject to the test rule" (ABA 2006, *18*). The chief problems, once again, stem from information scarcity and scientific uncertainty. Because variations in size, shape, surface characteristics, composition, and reactivity interact to affect behavior, broad generalizations are difficult. The information necessary to make a credible risk-based finding simply may not exist.

QUESTIONS ABOUT INSTRUMENT SELECTION

What tools do we use? Nanotechnology provides an interesting puzzle for regulatory design. We have already considered several questions leading to some disconcerting conclusions. First, it is uncertain whether the risk assessment process that provides the analytical framework for existing regulations is applicable to nanotechnology. Second, it is questionable whether regulators can manage the problems of model uncertainty, given the underlying complexity of the science and the dearth of research on health and environmental consequences. Third, it is debatable whether the regulatory authority conveyed by TSCA provides EPA with the means of leveraging information and, reflecting the lack of scientific research, mustering the scientific evidence that a given nanoscale version of a chemical currently in the inventory constitutes a reasonable risk.

But let us assume, for the sake of argument, that these formidable barriers could be surmounted and that EPA could arrive at defensible exposure thresholds. What then? Following Coglianese and Lazer (2003), we can conceptualize a key regulatory design decision—technology-based versus performance-based standards—by considering two questions: how homogeneous are the entities to be regulated, and how great is the regulator's capacity to assess outputs? Consider the first question. When high levels of homogeneity characterize regulated industries, the informational and resource demands of designing technology-based standards are minimized. A "one-size-fits-all" solution may be technically possible. As heterogeneity grows, the technical demands of developing standards increase until, at high levels of heterogeneity, standards might have to be customized for each facility. Fortunately, if the regulatory capability to assess outputs is high, one can make a strong case for performance-based standards wherein regulated entities are given maximum levels of compliance flexibility but held accountable for the results. When both conditions prevail, with high levels of homogeneity and high capacity to assess outputs, both approaches can be justified, although the decision may ultimately be shaped by cost or political considerations.

Nanotechnology is best understood as an emerging technological paradigm with applications that cut across industries ranging from cosmetics to information technology. Even if we assumed that the scientific foundations of regulation were in place, the maximal levels of heterogeneity would undermine the justification for technology-based standards. What of the capacity to assess outputs? Even if regulators could mount the scientific foundations for policy, it is doubtful that cost-effective technologies could be deployed at present to detect nanomaterials in environmental samples. Although some policies do in fact require the measurement of small particles—for example, the National Ambient Air Quality Standard for particulate matter—this measurement technology is inapplicable at the nanoscale. As EPA (2005, 43) notes, these technologies "were designed to effectively function on micron sized particles, particles hundreds to a thousand times larger than nanoparticles. Many of these technologies are not effective in the separation or analysis of particles at the nanometer scale." J. Clarence Davies (2006, 14) clearly expresses the regulatory implications of this statement: "If these materials cannot be detected, the provisions of the environmental laws are inoperable."

In sum, regulatory design presents an interesting puzzle. EPA needs to regulate a class of substances surrounded by uncertainty with respect to the underlying science, with little or no long-term experience or toxicological data for assessing impacts, few statutory instruments for filling the informational gaps, and limited technological means, short of sophisticated laboratory equipment, for monitoring levels of exposure (Wilson 2006). The high levels of heterogeneity among potential objects of regulation combined with the incapacity to assess outputs would seem to militate against both technology-based and performance-based standards.

QUESTIONS ABOUT SELF-REGULATION

Given such uncertainty, is self-regulation a reasonable option? Although many speak of self-regulation, little consensus exists as to what the term means. Is it an honest means of extending regulatory capacity or a neo-liberal smokescreen used to veil regulatory retrenchment? In its most minimalist expression, self-regulation can be little more than an exercise in public relations, with only minor impacts on corporate practices. Yet, as Cary Coglianese explores more fully in Chapter 4, when most scholars of regulation speak of self-regulation, they are addressing something far more significant. A firm may engage in voluntary self-regulation as a means of managing potential liabilities and its reputation among key stakeholders, such as customers, investors, or communities. It may introduce internal planning processes, rules, and protocols to manage environmental, health, or safety problems; it may monitor its behavior and take corrective action to bring its performance into alignment with larger organizational goals. To the extent that these activities reinforce the goals of public regulation, self-regulation should be welcomed by regulators.

Corporations may be required by their trade associations to exercise some self-regulatory functions. Associations have a strong interest in managing the reputation of industry actors and preempting new regulations that may be introduced in response to salient events. It is not surprising that events such as Three Mile Island, the Union Carbide chemical releases in Bhopal, and the Exxon *Valdez* oil spill were followed by significant expansions of trade association self-regulation. In some cases, associational self-regulation finds expression in detailed codes of conduct, protocols for environmental management systems, third-party inspections, and certification. If backed with significant sanctions, such as expulsion from the association or a loss of access to industry supply chains, self-regulation can serve an important role in governing corporate activity.

In its maximal expression, self-regulation, at the level of either the firm or the association, may be integrated with public regulation. Under "mandated full self-regulation" (Gunningham and Rees 1997) or "enforced self-regula-tion" (Braithwaite 1982), regulators may mandate that companies or associations write their own rules, subject to agency approval, and enforce them under conditions of regulatory oversight. Although regulators do not dictate the rules applied within the organization, they assume an oversight role and either engage in periodic inspections or employ third-party auditors. In essence, regulators make corporations function as "regulatory surrogates" that "provide considerable opportunities to extend the means of social control" (Gunningham et al. 1999, 222). In the United States, elements of self-regulation have been integrated with public regulation in various professions, such as medicine and law; securities (McCraw 1982); and more recently, food safety (Antle 1996), nuclear energy (Barkenbus 1983; Rees 1994), and environmental protection and occupational safety and health (Coglianese and Lazer 2003).

Self-regulation can be justified for myriad reasons. From a normative perspective, the best regulatory strategy would be one that stimulates the development of an industrial morality: "a form of moral discourse capable of challenging conventional industrial practices ... including the economic assumptions underlying many of those taken-for-granted policies and practices" (Gunningham and Rees 1997, 376), thereby reducing the need for regulation. From a political perspective, devolving functions can reduce the constraints imposed by stagnant regulatory budgets and political impasses that have prevented the expansion of statutory authority. Most important for present purposes, by assigning responsibilities to those with the best and most up-to-date information about core technologies, products, and processes, self-regulation can mitigate the informational asymmetries that set a hard limit on what can be accomplished through standard forms of public regulation.

Therefore, although the term "self-regulation" may suggest a withdrawal of public regulation, in its maximal expression it "represents not a reversal of twentieth century trends to nineteenth century self-regulation but an embedding of self-regulation within the late twentieth and early twenty-first century regulatory state" (Bartle and Vass 2007, 901–902). Indeed, to the extent that self-regulation and public regulation are integrated, it may be more appropriate to speak of coregulation. The conditions under which this embedding occurs are critically important. Thus much of the recent research has explored the ways public policy can be used to facilitate effective self-regulation. In the area of environmental protection, for example, "green tracks" have been employed as part of a regulatory pyramid in which firms with a verifiable record of regulatory compliance, a quality environmental management system, and third-party auditing are granted greater discretionary authority.

QUESTIONS ABOUT THE POTENTIAL FOR SELF-REGULATION IN NANOTECHNOLOGY

To what extent is self-regulation appropriate for nanotechnology in particular? As an emerging technological paradigm with applications that cut across industries, firms involved in the commercialization of nanotechnology are likely to be characterized by maximal levels of heterogeneity. Moreover, in light of the sheer complexity of nanotechnology, the dearth of scientific research on health and environmental effects, and the lack of cost-effective monitoring technologies, the regulatory capacity to assess outputs is non-existent. Given the defining features of nanotechnology as it exists today, the standard regulatory design question—technology-based or performance-based standards—is largely irrelevant. Following Coglianese and Lazer (2003), when regulated industries are characterized by high levels of hetero-geneity and regulators have a limited capacity to assess outputs, a strong case can be made for management-based regulation at the level of the firm or that of the industry via associations.

Yet one might argue that even management-based regulation would be premature at this point. Too little research has been conducted on the environmental, health, and safety consequences of nanotechnology, and such research by necessity must precede management-based regulation, as regulators and corporate managers must have a sense of what problems exist and how they can be prevented before they can design meaningful internal planning processes and protocols. Economical technologies for measurement also have to be in place before firms can monitor their own behavior (Coglianese and Nash 2001). But assuming that research and technologies will be forthcoming, a regulatory design that recognizes the importance of management-based regulation makes great sense.

Standards are currently in development under the auspices of the International Organization for Standardization (ISO), the long-established international consortium of national standards-setting agencies, including the National Institute of Standards and Technology in the United States. The Technical Committee 229, formed in 2005, consists of technical advisory groups from 30 participating nations and 9 observing nations and has formal liaison relationships with the Organisation for Economic Co-operation and Development (OECD). It is currently working to develop standards for terminology and nomenclature in nanotechnology (ISO 2005). Ultimately, it will develop standards for test methodologies, instrumentation, and scientifically informed environmental, health, and safety practices. If the current ISO standards provide a guide to the future, we can anticipate that ISO will integrate the control of nanotechnology into an environmental management system and require third-party auditing for certification. Past ISO standards for quality control (ISO 9001) and environmental management systems (ISO 14001) were rapidly adopted and became necessary for firms wishing to access many corporate supply chains and government bid lists. If a similar pattern of dissemination emerges for nanotechnology, self-regulation may progress far faster than the development of governmental regulatory capacity.

QUESTIONS ABOUT PARTNERSHIPS

What kind of collaboration is required? Industry standards can provide a foundation for self-regulation, but only if they are backed with scientific research on the environmental, health, and safety impacts of nanotech-nology. How can research be accelerated and the collection of information promoted more generally? At present, beyond governmentally supported programs such as the new EPA–NSF Centers for Environmental Implications of Nanotechnology, the funds devoted to such research are trivial relative to the investments in commercial development. Moreover, we have no reason to believe that industry-generated discoveries regarding potential environmental, health, and safety risks will be shared voluntarily with regulators. When EPA created the High Performance Voluntary (HPV) Challenge Program, a voluntary partnership with the chemical industry to promote the development and submission of toxicological information on high production volume chemicals, it discovered that 59 percent of the information (6,800 studies) was already in the possession of firms but had not been made publicly available. Under the regulatory authority granted by TSCA, no mechanism existed by which regulators could leverage information that companies often protected as proprietary or "trade secrets" (EPA 2004b).

Beyond direct funding, how can research of interest to regulators be promoted? Moreover, how can relevant information be collected that may prove consequential in understanding the regulatory implications of nano-technology? As explained above, TSCA is likely to be of little use in this endeavor. Yet other possibilities do exist. Governments in several countries have invested heavily in nanotechnology, and access to this funding might be made contingent on a commitment to devote some proportion of research to environmental, health, and safety issues and make the results available to regulators. Access to patent protection or government contracts could be contingent on the provision of information. Regulators might suggest that, in the absence of the voluntary provision of research findings, a regime comparable to that established under the Federal Insecticide, Fungicide, and Rodenticide Act (FIFRA) would be enacted. Under FIFRA, firms must make the affirmative case that their products will not cause harm when used as intended. The burden of proof on the agency as established in TSCA is reversed, with the onus on the corporation seeking regulatory approval.

Such efforts, if made unilaterally, would likely have little effect. In a global economy with high levels of investment and competition over nanotechnology, incentives for free-riding will be great. Fortunately, international coordination over the future of nanotechnology is occurring (in the form of the OECD's Working Party on Nanotechnology, for example), and a coordinated effort to leverage and share information could be quite important in filling the enormous gaps and generating the data necessary to inform future regulatory design and help develop the key features of management-based regulation at the level of the firm.

QUESTIONS ABOUT THE FUTURE OF REGULATION

Should the old regulations be adapted or new ones created? Certainly, a strong argument can be made for management-based regulation. Will such regulations be integrated into a broader system of public regulation? Will they be combined with mandatory technology-based or performance-based standards? At present, the information necessary to arrive at an answer is lacking. One could imagine a scenario in which the environmental, health, or safety ramifications of exposure to specific nanosize particles or structures would be so significant as to warrant mandatory controls. Should that occur, a strong case could exist for regulatory pluralism, combining mandatory controls and elements of firm- or association-based self-regulation.

Yet in the past, EPA has had difficulties integrating various forms of self-regulation into its existing regulatory structure. Take the example of ISO 14001, which establishes international standards for company-based environmental management systems. Although it participated heavily in the development of these standards, and despite the obvious benefits of certification, EPA to date has failed to recognize 14001 as the standard of choice. For example, between 2000 and 2009, EPA experimented with a now discontinued National Environmental Performance Track that provided greater flexibility to firms dedicated to "continuous self-improvement." Although it required that participants employ environmental management systems, it never gave privileged status to firms with ISO 14001 certification, nor did it provide firms more generally with regulatory incentives for self-regulation, such as a lower inspection priority (Coglianese and Nash 2001; EPA 2009).

As Marc Landy explores more fully in Chapter 5, the fault here lies less with EPA as an agency than with its legal room to maneuver. That is, the exhaustive regulatory statutes that provide a foundation for environmental regulation, most of them enacted in the 1970s, give agency officials little discretionary authority, thereby creating serious barriers to flexibility or reform. These gaps in existing regulatory authority have been partially filled by various partnerships and voluntary initiatives like Performance Track, but the results often have been difficult to verify or less than satisfactory (Eisner 2007).

Given the experiences to date, it would be unfortunate if the eventual regulation of nanotechnology were grafted on to existing regulatory capacity. Doing so would seriously reduce the extent to which regulatory design is informed by an understanding of the unique features of this emerging technology and the benefits of alternative approaches to regulation.

REFERENCES

ABA (American Bar Association). 2006. *Regulation of Nanoscale Materials under the Toxic Substances Control Act.* Chicago, IL: American Bar Association, Section of Environment, Energy, and Resources.

Andrews, Richard N. L. 2006. Risk-Based Decision Making: Policy, Science, and Politics. In *Environmental Policy: New Directions for the Twenty-First Century,* edited by N. J. Vig and M. E. Kraft. Washington, DC: CQ Press, 223–49.

Antle, John M. 1996. Efficient Food Safety Regulation in the Food Manufacturing Sector. *American Journal of Agricultural Economics* 78 (5): 1242–47.

Balleisen, Edward J., and Marc Eisner. 2009. The Promise and Pitfalls of Co-Regulation: How Governments Can Draw on Private Governance for Public Purposes. In *New Perspectives on Regulation,* edited by D. Moss and J. Cisternino. Cambridge, MA: The Tobin Project, 127–49.

Barkenbus, Jack N. 1983. Is Self-Regulation Possible? *Journal of Policy Analysis and Management* 2 (4): 576–88.

Bartle, Ian, and Peter Vass. 2007. Self-Regulation within the Regulatory State: Towards a New Regulatory Paradigm? *Public Administration* 85 (4): 885–905.

Bass, Carole. 2008. As Nanotech's Promise Grows, Will Puny Particles Present Big Health Problems? *Scientific American*, February 15.

Baumgartner, Frank R., and Bryan D. Jones .1993. *Agendas and Instability in American Politics*. Chicago, IL: University of Chicago Press.

Bergeson, Lynn L., Lisa M. Campbell, and Lisa Rothenberg. 2000. TSCA and the Future of Chemical Regulation. *EPA Administrative Law Reporter* 15 (4): 1–23.

Bosso, Christopher. 1987. *Pesticides and Politics: The Life Cycle of a Public Issue*. Pittsburgh, PA: University of Pittsburgh Press.

Braithwaite, John. 1982. Enforced Self-Regulation: A New Strategy for Corporate Crime Control. *Michigan Law Review* 80 (7): 1466–507.

Coglianese, Cary, and David Lazer. 2003. Management-Based Regulation: Prescribing Private Management to Achieve Public Goals. *Law & Society Review* 37 (4): 691–730.

Coglianese, Cary, and Jennifer Nash. 2001. Environmental Management Systems and the New Policy Agenda. In *Regulating from the Inside: Can Environmental Management Systems Achieve Policy Goals?*, edited by Cary Coglianese and Jennifer Nash. Washington, DC: Johns Hopkins University Press/RFF Press, pp1–26.

Davies, J. Clarence. 2006. *Managing the Effects of Nanotechnology*. Washington, DC: Woodrow Wilson Center for Scholars.

———. 2007. *EPA and Nanotechnology: Oversight for the 21st Century*. Washington, DC: Woodrow Wilson Center for Scholars.

Denison, Richard A. 2009. Ten Essential Elements of TSCA Reform. *Environmental Law Reporter* 39: 10022–28.

Dosi, Giovanni. 1982. Technological Paradigms and Technological Trajectories. *Research Policy* 11:147–62.

EDF (Environmental Defense Fund). 1997. *Toxic Ignorance: The Continuing Absence of Basic Health Testing for Top Selling Chemicals in the United States*. New York, NY: Environmental Defense Fund.

Eisner, Marc Allen. 2007. *Governing the Environment: The Transformation of Environmental Protection*. Boulder, CO: Lynne Rienner Publishers.

EPA (U.S. Environmental Protection Agency). 1998. *Chemical Hazard Data Availability Study: What Do We Really Know About the Safety of High Production Volume Chemicals?* Washington, DC: U.S. EPA.

———. 2004a. *An Examination of EPA Risk Assessment Principles and Practices*. Washington, DC: U.S. EPA.

———. 2004b. *Status and Future Directions of the High Production Volume Challenge Program*. Washington, DC: U.S. EPA.

———. 2005. *Nanotechnology White Paper. External Review Draft*. Washington, DC: U.S. EPA.

———. 2007. *Overview: Office of Pollution Prevention and Toxics Programs.* Washington, DC: U.S. EPA.

———. 2009. Enforcement and Compliance Guidance on the Termination of the National Environmental Performance Track Program. www.epa.gov/perftrac/downloads/OECA-09-000-9690memo.pdf (accessed November 25, 2009).

Gunningham, Neil, Martin Phillipson, and Peter Grabosky. 1999. Harnessing Third Parties as Surrogate Regulators: Achieving Environmental Outcomes by Alternative Means. *Business Strategy and the Environment* 8: 211–24.

Gunningham, Neil, and Joseph Rees. 1997. Industry Self-Regulation: An Institutional Perspective. *Law & Policy* 19 (4): 363–414.

ISO (International Organization for Standardization). 2005. TC 229: Nanotechnologies. www.iso.org/iso/iso_technical_committee?commid=381983 (accessed February 28, 2009).

———. 2007. *Business Plan: ISO/TC 229 Nanotechnologies.* Geneva, Switzerland: International Organization for Standardization.

McCraw, Thomas K. 1982. With Consent of the Governed: SEC's Formative Years. *Journal of Policy Analysis and Management* 1 (3): 346–70.

NMAB (National Materials Advisory Board). 2006. *A Matter of Size: Triennial Review of the National Nanotechnology Initiative.* Washington, DC: Committee to Review the National Nanotechnology Initiative, National Materials Advisory Board, Division on Engineering and Physical Sciences.

NNI (National Nanotechnology Initiative). No date. So What Is Nanotechnology? www.nano.gov/html/facts/whatIsNano.html (accessed February 28, 2009).

———. 2009. Research and Development Leading to a Revolution in Technology and Industry: Supplement to the President's FY2010 budget. www.nano.gov/NNI_2010_budget_supplement.pdf (accessed November 25, 2009).

NSTC (National Science and Technology Council). 2007. *The National Nanotechnology Initiative Strategic Plan.* Arlington, VA: National Nanotechnology Coordination Office.

Rees, Joseph V. 1994. *Hostages of Each Other: The Transformation of Nuclear Safety Since Three Mile Island.* Chicago, IL: University of Chicago Press.

Rhomberg, Lorenz R. 1996. *A Survey of Methods for Chemical Health Risk Assessment among Federal Regulatory Agencies.* Washington, DC: National Commission on Risk Assessment and Risk Management.

Ruckelshaus, William D. 1983. Science, Risk, and Public Policy. *Science* 221 (4615): 1026–28.

Tolstoshev, Arius. 2006. *Nanotechnology: Assessing the Environmental Risks for Australia.* Melbourne, Australia: Earth Policy Center.

Wilson, Robin Fretwell. 2006. Nanotechnology: The Challenge of Regulating Known Unknowns. *Journal of Law, Medicine & Ethics* 34 (4): 704–13.

CHAPTER 4

ENGAGING BUSINESS IN THE REGULATION OF NANOTECHNOLOGY

Cary Coglianese

Nanotechnology holds great promise but also raises concerns about possible health and environmental risks. To date, the fears about nanotechnology have far outstripped what is actually known about the health benefits and costs of this technology. The lack of information about the health and environmental effects of nanotechnology poses distinct challenges about whether and how to regulate the manufacturing and use of engineered nanomaterials (Sylvester et al. 2008). In at least this one respect, though, nanotechnology is not novel, because regulators in almost all contexts find themselves needing more information than they have—and regulators are consistently at a comparative informational disadvantage vis-à-vis the business firms they regulate (Coglianese 2007).

One solution to the general information asymmetry facing regulators has been to engage business in the process of regulating, whether by adopting participatory decisionmaking procedures, selecting regulatory instruments that give firms flexibility or otherwise delegating regulatory responsibility to them, or encouraging the development of voluntary efforts by businesses. Each of these efforts to engage business aims to take advantage of industry's informational superiority in developing effective solutions to environmental, health, and safety problems. But in having regulators take advantage of industry, it is also possible for industry to take advantage of regulators and, by extension, the public. After all, if regulation is needed, this is because the interests of business are not fully aligned with the broader interests of the public. As a result, efforts to engage business in regulating business present inherent challenges. Some of these challenges are those that exist any time regulators seek to alter the behavior of business; however, these challenges surely prove more salient the more that businesses receive flexibility or assume their own regulatory responsibilities.

This chapter reviews what researchers have learned about engaging business in the process of regulation, including what is known about achieving governance without the involvement of government. Extrapolating from this research literature, it appears that the current lack of information about the full social effects of nanotechnology creates a strong case for engaging with and even delegating to industry, but this same lack of information also makes it difficult to ensure that businesses will respond to such efforts to engage them by acting in a socially optimal manner. In the end, the forms of engagement or delegation that may appear to be attractive in light of the underlying information deficit about nanotechnology will likely face inherent difficulty in ensuring effectiveness in regulating this new technology—or at least it will be difficult for regulators or the public to know whether these approaches are effective.

CONDITIONS FOR REGULATING NANOTECHNOLOGY

As Marc Eisner suggested in Chapter 3, some existing environmental and safety regulations arguably already apply to nanotechnology (Davies 2006). Yet even if they do not, the kinds of concerns raised about nanotechnology meet the conventional market-failure test for new regulation. This test assumes that competitive markets generally prove successful for producing and allocating society's resources, in part because they provide incentives for innovations such as nanotechnology, but it also recognizes that conditions for socially optimal free-market transactions are not always obtained (OMB 2000; Stokey and Zeckhauser 1978). When these conditions are not met, market failures arise.

Market failures typically fall into three categories. First, the market fails when competition is lacking, as in cases of monopoly. Second, market failure arises from externalities—that is, from the failure of market prices to incorporate all the costs to society of particular economic behavior. Environmental regulation responds to this type of market failure, seeking to alter firms' behavior in ways that reduce negative externalities. Third, markets can fail when price or wage signals do not match what they would be if contracting parties had full information about products and services. Government regulation is justified to overcome these kinds of information asymmetries, whether through product bans or standards or through various kinds of labeling and product-testing requirements.

Nanotechnology could conceivably implicate any of these three categories of market failure. Although known instances of harm from nanomaterials are virtually nonexistent—indeed, as recent as only a few years ago it could be said that "there have been no known cases of people or the environment being harmed by nanomaterials" (Davies 2007, 14)—red flags have certainly been raised about the potential health effects of nanotechnology (Breggin and Carothers 2006). This chapter proceeds from the assumption that the most salient concerns about nanotechnology arise over externalities and information asymmetries. Engineered nanomaterials that threaten human health could potentially create externalities when introduced into the environment, whether as waste products during production or after consumption. These same materials also may pose hazards to workers and possibly consumers as well, even though workers and consumers are not likely to understand the full risks.

Given the reasonable likelihood of some need for a regulatory response to certain nanotechnologies, this chapter explicates what we know about the role business can play as part of a strategy aimed at addressing public regulatory problems. Rather than defend any particular regulatory response to nanotechnology, I seek primarily to inform future decisionmaking about how to respond to nanotechnology by assessing what is known more generally about different strategies of engaging business in the process of regulation, with or without government involvement. Any effort to learn from experience in other realms will be relevant only by keeping the context of nanotechnology in mind, in particular the vast uncertainty about whether or how some kinds of engineered nanomaterials may be harmful to health and the environment (Mandel 2008). As with the other chapters in this volume, the discussion here is based on the following additional assumptions about nanotechnology:

- Nanotechnology is not a single technology, but refers to a host of different types of man-made particles that have in common little more than their tiny size.

- Even if some engineered nanomaterials pose the risk of negative health or environmental effects, this does not mean that all nanomaterials do.
- Some nanomaterials may help improve public health, whether through the development of better medications, safer materials, or new methods of monitoring or cleaning up environmental conditions.
- The "nanotechnology industry" is highly heterogeneous, as nanomaterials are being developed and used by businesses across a variety of preexisting industrial sectors, from cosmetics to chemicals, pharmaceuticals to petroleum.
- The processes used in manufacturing nanomaterials, as well as their uses in the manufacturing of other products, are also highly heterogeneous.
- Risk assessment of nanotechnology remains in its infancy, and efforts to understand the health effects of nanotechnology are likely in the near to medium term to be outpaced by rapid development of new nanomaterials and new applications of nanotechnology.

Given these assumptions, what can we discover about various options available for engaging business in the regulation of nanotechnology? The first question is: what are the various options? The next section addresses this question by presenting a framework for regulatory choice that also provides the basic organizing frame for the remainder of the chapter.

CHOICE OF REGULATORY APPROACHES

To say that regulating nanotechnology might be justified because of externalities or information asymmetries is not to say that *any* kind of regulation is appropriate. Even if some regulatory tool is needed, that certainly does not mean that even the most blunt or ill-suited instrument should be pulled out of the toolbox. Regulatory approaches or instruments vary greatly. For example, Richards (2000) summarizes more than a dozen different taxonomies, each containing about six or seven different labels used to describe discrete policy instruments in the field of environmental regulation. Examples include technology requirements, performance standards, emissions taxes, and pollutant trading systems.

Although the array of instruments available to any decisionmaker may seem large, regulatory tools all share common attributes. All regulatory instruments consist of some rule or rule-like statement having normative force and backed up with some type of consequences. Given these core similarities, the differences between the myriad regulatory instruments can be explained in terms of four variables: the regulator, the target, the type of command, and the type of consequences.

1. Regulator. The entity that creates the rule and dispenses the consequences is the regulator. It is possible for the rule creator to be different from the rule enforcer, but usually these are one and the same. The regulator is typically thought to be a legislature or governmental agency, but such an entity can take the form of various nongovernmental standard-setting bodies (such as the International Organization for Standardization), nonprofit organizations (such as Underwriters Laboratories), industry trade associations (such as the American Chemistry Council), or even business firms themselves when they impose rules on their employees. The distinction between regulation and self-regulation is simply based on who the regulator is (Sinclair 1997). Just as with a government regulation, an industry regulator can adopt rigid technology requirements or deploy more flexible performance- or market-based standards.

2. Target. The regulated entity is the individual or organization to which a regulatory instrument applies and on whom or which consequences can be imposed. Usually this entity is also the principal factual trigger or frame of reference for the regulation. But that trigger can be smaller or larger. For example, if an air-pollution regulation prohibits industrial facilities from emitting pollution from any smokestack above a specified level, the target is still the individual facility, even though the trigger is an individual smokestack. By contrast, an air-pollution regulation which has an entire facility as its trigger or frame of reference would thereby allow regulated facilities to vary emissions across different smokestacks, so long as average emissions from each facility do not exceed a specified level. The frame of reference can be broader still with full-blown emissions-trading regimes, in which case the entire sector (say, all coal-powered utility plants in the Midwest) can be targeted with an overall emissions reduction, but individual facilities can sell or trade emissions permits. With emissions trading, an individual facility is still the regulated entity in the formal sense that it is subject to the rule ("emit no more pollution than you have permits for"), but the sector is the explicit frame of reference for the regulatory regime.

3. Command. A rule can direct that a target adopts means or achieves ends. In other words, it can direct the target to engage in or avoid a specific action designed to advance the regulatory goal, such as a command to install

ventilation systems or provide employees with protective equipment, or it can compel the target to achieve or avoid a specified outcome related to the regulatory goal, such as a rule stating that emissions shall not exceed a specified level or that workplaces shall not have levels of contaminants in the air exceeding a certain concentration level. In addition, regulation can command the disclosure of information, which can be viewed as either a particular kind of means, such as when disclosure is used to create consumer or shareholder pressure for a target to achieve a desired end, or as the end itself, such as when the regulator seeks the end state of information availability to consumers. Finally, regulatory commands can leave the choice of means and ends to the target, instead requiring it to plan and develop its own internal set of rules aimed at addressing a regulatory problem.

4. Consequences. The normative force of any command must be reinforced with consequences. Consequences can be negative in the form of penalties, such as fines or the loss of a license, or positive in the form of product approvals, regulatory exemptions, or other rewards granted once a target meets the predicate conditions in a rule. Consequences can also be distinguished by what might be considered their functional form. Often consequences take a binary form—in other words, if a rule is violated, a lump-sum penalty is issued. With such a binary consequence, it does not matter whether the rule is violated by a small or large degree; the penalty is the same. But consequences can also be applied on an incremental or marginal basis. For example, emissions tax schemes vary the consequences incrementally: for every additional unit of pollution emitted, the target pays a corresponding additional unit of money.

Implications

These major regulatory variables—regulator, target, command, and consequences—affect the degree to which business has regulatory flexibility and responsibility. Business has maximal flexibility and responsibility when the regulator is the industry or corporation itself. Yet even when the regulator is the government, choices made about variables such as the target and command will affect how business engages in decisionmaking to address a regulatory problem. A command that dictates achievement of an end state allows firms to determine how to achieve that dictated end state—and the closer that commanded end state is to the motivating purpose or ultimate end state of a performance standard, the greater the flexibility it will generally afford. A command that merely dictates planning, and allows firms to choose specific actions or objectives, gives firms still more flexibility.

The sections that follow consider various approaches to involving business in regulation and explore what can be learned about these approaches from contexts other than nanotechnology. I will begin with approaches that involve the government as the regulator, and then move to nongovernmental approaches. Three caveats or limitations are in order.

First, even when focused on governmental approaches, I am concerned here with the form of regulation and not particular governmental institutions, such as the Food and Drug Administration (FDA) versus the Environmental Protection Agency (EPA). Although some existing governmental agencies may happen to use one type of regulatory command more than others, and although it is useful to learn from those agencies' experiences, my principal aim is to draw out lessons about the design of rules rather than about particular institutions. By contrast, in Chapter 5, Marc Landy looks specifically at the capacity of and approaches taken by EPA as a regulator in addressing the potential environmental effects of engineered nanomaterials.

Second, in discussing different approaches, I largely leave aside questions about the stringency of regulatory standards (that is, whether a particular performance standard is set at a demanding level). This is not because stringency is not important, but largely because what we know about regulatory instruments comes from studies of their general form; to date, researchers have not factored stringency systematically into comparisons of different instrument types.

Finally, in assessing regulatory approaches, I give primary attention to the effectiveness of different approaches in achieving the principal regulatory goal, such as risk reduction. Other criteria, such as cost-effectiveness, are certainly relevant and should be considered by regulators before selecting any particular approach. But my premise here is that approaches that extensively involve business in regulation will inherently tend to fare well in terms of cost-effectiveness. Because minimizing costs is in business's self-interest, firms will be better able to advance cost-effectiveness the greater flexibility or responsibility they have. The central question is whether methods of engaging business can also sufficiently advance the public's regulatory goals, even if these might not be a priority for firms in terms of their immediate self-interest.

FLEXIBLE FORMS OF GOVERNMENT REGULATION

Certain kinds of regulation, even when imposed by the government, can require engagement by industry in addressing public problems. The usual image of government regulations is that they do not allow business to engage in finding solutions to public problems. This image of so-called command and control regulation especially applies when regulations dictate that businesses adopt specific means of achieving social goals. Such means-based regulations, commonly called technology, design, or specification standards, sometimes require adoption of a specific technology, such as when the Federal Aviation Administration required airlines to install locks on the doors of airplane cabins to prevent hijacking. Other times they prohibit certain actions, such as selling toys coated with lead paint. In either case, businesses are given little flexibility in choosing how to act. Although they may explore taking other actions that also advance the public goal, businesses will have no choice but either to comply with the specific measures that government dictates—or suffer the consequences.

Means-based regulation is effective when the mandated means actually contribute to achieving the public policy goal when applied by different regulated firms. The government regulator must understand enough about how relevant business operations contribute to the policy problem, and about options for ameliorating that problem, in order to select an option that will remedy the problem. If the causes of the problem are complex or not well understood, or if business operations vary widely across different firms, then mandating one particular means will prove to be ineffective, costly, or even counterproductive. In addition, by mandating adoption of a specific course of action, firms will have less incentive to seek out new methods of solving the problem that could be more effective or less costly (Gunningham and Johnstone 1999).

In contrast to means standards, government has available three alternative regulatory instruments that give firms more flexibility: performance standards, information disclosure requirements, and management-based regulation. An advantage of these flexible types of regulation is that government needs less information about the specific actions necessary to advance environmental quality, worker safety, consumer protection, or other policy goals.

Performance Standards

A performance standard specifies the level of performance required of a firm without specifying how the firm will achieve that goal (Viscusi 1983). For example, a regulation may limit the exposure of workers to particular hazardous chemicals but not specify how exposure levels are to be achieved. Such an approach gives firms the flexibility to find less costly ways to achieve the required performance levels (Gunningham 1996).

Performance standards applied uniformly across firms can be overinclusive, however, because they require all firms to control to the same level, even if they have different compliance costs (Hahn 1989). Small businesses in particular chafe at the imposition of sectorwide standards that are met more easily by their larger counterparts and argue that such standards only reinforce market biases that favor larger businesses. A more cost-effective approach gives firms with lower marginal compliance costs incentives to achieve higher-than-average performance levels, making up for lower performance by firms facing higher costs. By allowing for nonuniform levels of performance, market-based performance standards can reduce overall costs and provide still greater incentives for firms to innovate (Ackerman and Stewart 1985; Hahn and Hester 1989; Pildes and Sunstein 1995; Stavins 1989, 2003; Tietenberg 1990).

The flexibility afforded by performance standards has made them attractive to policymakers. For instance, the executive order on regulation in existence from the Clinton administration to the Obama administration directs federal agencies wherever feasible to specify performance objectives, rather than mandate behavior, in crafting new regulations. In addition to the cost-saving potential of performance standards, these standards can also accommodate technological change in ways that prescriptive means-based standards generally cannot.

To work effectively, performance standards require that the regulator be able to monitor and assess the regulated outputs. This is not always possible, however, and performance-based standards hold fewer advantages when it is difficult or prohibitively expensive to assess critical outputs (Ayres and Braithwaite 1992). Assessing outputs will be difficult when a problem is poorly understood and it is hard to define performance in a precise way or to know at what level a performance standard ought to be set. Setting optimal performance levels also often requires a detailed understanding of the dose–response relationships between a problem's precursors and the ultimate outcome of concern. For example, setting optimal emissions thresholds requires an understanding of the relationship between emissions and human health outcomes. When these dose–response relationships are poorly understood, it is difficult to determine where performance levels should be set. Further difficulties arise when there is uncertainty about what precursors or proxies to embed in a rule.

Performance standards also face difficulties when monitoring technologies do not exist to measure performance in a reliable manner, or when it is costly for the regulator to measure and observe performance across a number of firms. In some cases, regulators end up relying on the regulated firms to conduct the monitoring and report results on a regular basis. Such delegated monitoring typically requires the existence of reliable measurement methods as well as a robust auditing function by the regulator. Where performance standards are both imprecise (such as the FDA's "safe and effective" standard for drug approvals) and dependent on firms for self-monitoring, this inevitably leaves room for interpretation, and the risk exists that regulated entities will report biased results. Even if they are acting in good faith, regulated entities are inclined to present or interpret their own models and data in ways that make it look as if their actions perform better than a more disinterested examination would find.

Information Disclosure

Another flexible regulatory approach is simply to require firms to gather information and disclose it to the government or the broader public, without otherwise requiring any change to firm behavior (Graham 2002; Karkkainen 2001; Kleindorfer and Orts 1998; Sunstein 1999). Information disclosure has been used in a wide range of regulatory areas: securities regulation (Ferrell 2007), auto safety (Graham 2002), and consumer and food products (Fung et al. 2007), to name a few. In the late 1990s, the county of Los Angeles compelled information disclosure to address the problem of cleanliness in restaurants. Under the county regulation, restaurants were required to display prominently outside their entrance a card disclosing the grade they received during their most recent health inspection. This requirement, simply to disclose an inspection grade, was found to be associated with an improvement in subsequent inspection scores and a 13 percent decline in the rate of local hospitalizations due to foodborne illnesses (Jin and Leslie 2003).

Clearly, information disclosure can be an effective strategy, at least under certain conditions. In the case of restaurant hygiene regulation, the conditions included the following: information disclosure was combined with other forms of regulation (that is, restaurant inspections were still conducted under traditional regulations, as these formed the basis of the hygiene grade); the information disclosed was readily available and easy for users to interpret (large cards displayed grades of A, B, or C); and the users had direct and clear interests that were activated by the information, and thereby disclosure shaped the incentives for the firm (restaurant-goers could walk down the street to a restaurant with a better score card).

Not all regulatory domains meet these conditions. For example, the federal Toxics Release Inventory (TRI) rule adopted in the late 1980s requires industrial facilities that use large volumes of toxic chemicals to disclose on an annual basis the amounts of toxins they release into the environment. Toxic releases in the United States subsequently declined by about 45 percent during the decade following the adoption of the TRI requirement. Researchers have attributed this decline to the disclosure regulation (Fung and O'Rourke 2000), reasoning that the disclosure effectively "shames" firms into improving their social performance (Graham 2002). For large emitters of toxic chemicals, the disclosure of pollution information is said to increase media scrutiny and community pressure, while causing an initial decline in stock prices (Hamilton 2005).

Nevertheless, any conclusion that the full decline in TRI releases can be causally attributed to the disclosure of TRI information is surely overstated. During the same period of time, the federal EPA implemented extensive conventional regulatory programs designed to address toxic pollutants, including new regulations under the Clean Air Act of 1990. In addition, a recent study has suggested that the reported decline in TRI emissions may be due in substantial part to an artifact in the program's paperwork requirements (Bennear 2008). As the most comprehensive study of TRI to date has concluded, "The separate and exact impacts that the provision of information has on toxic emissions are, to date, unknown" (Hamilton 2005, 242).

Management-Based Regulation

As with performance standards and information disclosure, management-based regulation allows firms the flexibility to choose their own control or prevention strategies (Coglianese and Lazer 2003). But management-based regulation is different from performance-based regulation in that it mandates specific, and sometimes extensive, planning and management activities. In this way, it represents more than a requirement just to generate and disclose information. Under management-based regulation, or what some have called "enforced self-regulation" (Bardach and Kagan 1982; Braithwaite 1982; Hutter 2001), firms are expected to produce plans that comply with general criteria designed to promote the targeted social goal. Regulatory criteria for planning specify elements that each plan should have, such as the identification of hazards, risk-mitigation actions, procedures for monitoring and correcting problems, employee-training policies, and measures for evaluating and refining the firm's management with respect to the stated objective. Plans sometimes are made subject to approval or ratification by government regulators. Similarly, under some management-based regulations, firms are required to produce documentation of subsequent compliance or are subjected to reviews by regulators or third-party auditors to determine compliance with mandatory planning rules.

Management-based regulation places responsibility for decisionmaking with business actors who generally possess the most information about risks and potential control methods. Thus, as with other flexible instruments, the actions that firms take under a management-based approach can be expected to be less costly and more effective than under government-imposed standards (Ayres and Braithwaite 1992). By allowing firms to make their own decisions, managers and employers are more likely to view their own organization's rules as reasonable; as a result, compliance may be greater than with government-imposed rules (Ayres and Braithwaite 1992; Coglianese and Nash 2001; Kleindorfer 1999). In this way, as well as by enlisting the assistance of private third-party certifiers, management-based regulatory strategies may help mitigate the problems associated with limited governmental enforcement resources. Finally, by giving firms flexibility to create their own regulatory strategies, management-based regulation enables firms to experiment and seek out better, more innovative solutions.

Management-based regulation has been implemented in a variety of areas, including food safety, chemical-accident avoidance, and pollution prevention (Bennear 2006). For example, following the deadly 1984 chemical accident at a Union Carbide plant in Bhopal, India, the U.S. Occupational Safety and Health Administration (OSHA) established standards for process safety management (PSM) of highly hazardous substances. OSHA's PSM regulation requires firms to implement a multistep management practice to assess risks of chemical accidents, develop procedures designed to reduce those risks, and take actions to ensure that procedures are carried out in practice. The core of the PSM protocol is a process hazard analysis. Firms must undergo an extensive analysis of what could potentially go wrong in their facilities' processes and what steps must be in place to prevent chemical accidents from occurring. OSHA defines "process" broadly to mean any use, storage, handling, or manufacture of toxic chemicals at a site. Each process must be analyzed separately, and then firms must rank each one according to factors such as how many workers could potentially be affected and its operating history, including any previous incidents involving the process. Firms must next identify both actual and potential interventions to reduce hazards associated with each process, including control technologies, monitoring devices, early-warning systems, training, or safety equipment.

Based on their analyses, firms must develop written operating procedures for both normal operating conditions and emergency situations. These procedures must be made available to employees who work with the chemical processes. In addition, OSHA requires that firms continuously review these procedures and update them as necessary to reflect process changes, new technologies, and new knowledge. Firms are required to certify their operating procedures annually and provide for compliance audits every three years. By such systematic tracking of process and incident data, firms are well positioned to make modifications that can improve worker safety (Chinander et al. 1998; Coglianese and Lazer 2003; Kleindorfer 2006). Although the precise impact of management-based regulation of chemical accidents is unclear, in part because accidents like the one at Bhopal are low-probability events to begin with, indications are that OSHA's PSM requirements, and similar requirements adopted by EPA, at least partly explain a 40 percent decline in damage claims in the chemical industry between 1987 and 1997 (Marsh & McLennan 1997).

Fourteen states have adopted similar toxic-related management-based regulations. The Massachusetts Toxic Use Reduction Act (TURA), for example, requires firms that use large quantities of toxic chemicals to analyze their use and flow throughout the facilities, develop plans to reduce the use and emissions of toxic chemicals, and submit reports on the plans to state environmental agencies (Karkkainen 2001). Massachusetts also requires that a state-authorized pollution-prevention planner certifies each plan. Although firms are required to go through the planning process and develop a system for reducing the use and emissions of toxic substances, TURA does not impose any specific performance requirements, nor even require that firms comply with their own plans. Moreover, firms' plans are considered proprietary and are therefore not made available to the public, thus eliminating the possibility of community pressures that publicly available plans could generate. The program aims to encourage firms to make gains in terms of pollution prevention simply by requiring them to go through a planning process.

Using careful statistical techniques, Bennear (2006) has systematically analyzed TRI releases from more than 30,000 facilities, comparing releases from facilities in the 14 states having programs similar to TURA with those in the remaining states that have no management-based regulation. She found that the presence of a management-based regulation in the jurisdiction within which a facility is located was associated with an overall decrease of about 30 percent in on-site TRI releases. She also tested for the effects on facilities over time. The most statistically significant effects (at the 5 percent level) occurred within two to four years after the imposition of a planning mandate; the statistical significance dropped for years five and six (10 percent level), and after year six, mandatory planning requirements showed no statistically significant effects on TRI releases. These results are consistent with interview data showing that following the imposition of management-based regulation, firms may initially "pick the low-hanging fruit" but over time tend to treat the planning requirements imposed by management-based regulation more like paperwork exercises (Coglianese and Nash 2004).

Prescribing a set of management processes thus does not necessarily guarantee that firms will have the motivation to use them to achieve socially optimal results. As with any regulatory regime, firms may simply go through the motions or game the system if they lack the motivation or incentive to use the planning process to achieve socially optimal benefits. Management-based approaches therefore still usually require a governmental enforcement presence to ensure that firms conduct the necessary planning and implement their plans if required. Because regulated firms can be expected to have interests at odds with the government's goals (by definition, this will be true if regulation is needed), and because firms have an informational advantage, the regulator must find ways to induce and control firms so that they manage themselves in ways more aligned with social goals. To the extent that firms lack adequate incentives on their own to create plans and implement them, they may resist—even in subtle ways—complying with the letter and spirit of any management-based mandate imposed by government. When it comes to planning, some firms undoubtedly will devote as few resources as they can to analysis. Others will have the incentive to produce plans that minimize implementation costs rather than maximize social benefits. The regulator therefore needs to be able to assess whether firms' planning processes and their resulting plans have been appropriately rigorous. When it comes to implementing plans, firms may have the incentive to avoid costly and effective implementation, so regulators must also have the capacity to assess whether a firm has made adequate capital investments and is regularly acting in a way consistent with its plans.

Regulators have several options to prevent shirking under management-based regulations. One is to develop specific—and more difficult-to-evade—mandates embedded within the management-based regulation. Another is to ensure the adequacy of planning by requiring government approval of firms' plans. To ensure effective implementation, regulators can impose suitably detailed record-keeping requirements and institute inspections or third-party audits.

Even with these measures, what makes for "good management" will necessarily remain somewhat open-ended or case-specific, and the greater discretion afforded to firms under a management-based approach to regulation will inevitably complicate enforcement. Government enforcement of management-based regulation can be difficult precisely because the same conditions that make such regulation attractive—namely, the limitations of means or performance standards—will also tend to make it more difficult for government to determine what constitutes responsible management.

GOVERNMENT VOLUNTARY PROGRAMS

Another way for government to engage business in addressing policy problems is to establish programs that try to recognize or reward businesses that voluntarily improve their performance. Many such programs have been adopted in the last two decades in the area of environmental policy. In recent years, the EPA has administered more than 60 so-called voluntary programs, and an even larger number of voluntary programs can be found at the state level. These programs include educational resources to firms, local governments, nongovernmental organizations, and citizens; financial support for projects that demonstrate beyond-compliance practices, including technical assistance programs; competitive awards recognizing firms that distinguish themselves beyond their peers for one-time achievement; product certifications that seek to promote "green" markets by developing product standards and processes for certifying that products meet these standards (EPA's well-known EnergyStar program being an example); and voluntary "partnerships" that involve an exchange between EPA and a regulated entity (the agency offers a package of benefits in exchange for certain types of beyond-compliance activities on the part of the facility).

Two prominent examples of such partnerships include EPA's 33/50 Program—the agency's earliest voluntary program—and the National Environmental Performance Track (NEPT), which had until recently been considered the agency's "flagship" voluntary program.

The 33/50 Program

EPA launched its 33/50 Program in early 1991, seeking to achieve the overall reduction of industrial releases of 17 target chemicals by 33 percent by the end of 1992 and 50 percent by 1995 (against a 1988 baseline). Any company that released any one of the designated chemicals was eligible to participate. To join, a company needed only to commit to reducing a minimum of one of these chemicals by any amount. Out of about 10,000 companies eligible for the program because they reported toxic releases, EPA invited close to 8,000 to participate, and about 1,300 eventually chose to join (EPA 1999; Khanna 2006).

Participating and nonparticipating companies achieved sufficient reductions in toxic releases to meet EPA's national 33 percent and 50 percent goals. EPA measured the achievement of these goals using the TRI. By the end of the first year of the program, the total reported releases of the 17 targeted chemicals were down more than 33 percent compared with 1988 levels. By the end of 1995, total releases of the targeted chemicals had dropped by 56 percent (831 million pounds) since 1988.

The 33/50 Program has been the subject of numerous reviews, which present a complex picture of its impacts (Khanna and Damon 1999; Gamper-Rabindran 2006; Sam and Innes 2004). No one seriously thinks that the program led to the entire 56 percent reduction in the release of targeted chemicals from 1988 to 1995. After all, 33/50 had not even been announced until early 1991, so it certainly was not responsible for the reductions that occurred before that time. Furthermore, TRI chemicals not included in the 33/50 Program also declined, suggesting that this would have happened with targeted chemicals as well.

Even EPA officials concede that some of the observed reductions in target releases occurred for reasons that had nothing to do with 33/50, including preexisting corporate pollution-control programs, the closure of a facility or elimination of a product line for economic reasons, the incentives provided by the public availability of TRI data, and other regulations on targeted chemicals. With respect to this last factor, it is worth noting that EPA issued a proposed or final rule directed at reducing almost all of the 17 target chemicals between 1988 and 1991. The 33/50 chemical with the single largest percentage reduction—1,1,1-trichloroethane, an ozone-depleting chemical found in industrial solvents—was subject to a regulatory ban under the 1987 Montreal Protocol. Reductions in this one chemical alone accounted for approximately 30 percent of the total reductions in 33/50 chemicals from 1990 to 1995. Using regression analysis, Khanna and Damon (1999) examined the effect of the program on firm-level releases in a sample of about 125 firms in the chemical industry, reporting that the 33/50 Program was associated with as much as a 28 percent decline in the target chemicals during the period 1991–1993. According to Morgenstern and Pizer (2007, 177), however, there is "some evidence that [this] effect is reversed when ozone-depleting substances [covered under the Montreal Protocol are] excluded." Sam and Innes (2004) concluded that the program corresponded with some reduction in releases and a decline in inspection rates. Gamper-Rabindran (2006) found that associated releases varied by sector, with releases in the fabricated metals and paper industries decreasing significantly, but releases in the chemicals and primary methods industries increasing significantly. More recent analyses of toxic releases from larger samples of firms across a range of sectors have continued to yield mixed results about 33/50's effectiveness (Vidovic and Khanna 2007; Innes and Sam 2008; Bi and Khanna 2008).

National Environmental Performance Track

EPA launched its National Environmental Performance Track (NEPT) program in 2000. Performance Track sought to induce facilities to make measurable environmental improvements by recognizing and encouraging top environmental performers. To qualify for membership in this voluntary program, a facility needed to have (a) implemented a formal, independently assessed environmental management system (EMS), (b) demonstrated a history of past environmental achievements and a record of sustained compliance with environmental regulations, (c) made commitments to improving its environmental performance; and (d) engaged in community outreach activities.

EPA's Performance Track staff would review applications for new members twice each year. Starting in 2000, membership in the program grew by about 10 percent annually, with about 440 facilities participating by

September 2007. Although members represented a wide range of economic activities, four major manufacturing sectors—chemical products, electronic and other electric equipment, pharmaceuticals and medical equipment, and transportation equipment—made up a substantial portion of the program's membership. Many nonmanufacturing facilities also participated, however, and in the program's later years a number of facilities from the arts, recreation, entertainment, research, education, and governmental sectors joined (EPA 2007a).

Once accepted into Performance Track, a facility's membership would be good for three years, at which point it needed to reapply. EPA expected members to make progress toward achieving their performance commitments, and the program required that facilities "normalize" their current performance vis-à-vis a baseline level. Each facility needed to complete an 11-page annual performance report (APR), signed by the senior manager responsible for the facility. EPA posted facility APRs on its website, and facility managers were expected to make the report generally available to the public.

According to EPA, the program's positive social externalities were significant, with the agency reporting that, between 2004 and 2005, members reduced total materials use by nearly 25,000 tons, water consumption by 1.7 billion gallons, energy use by 4.3 million BTU, and hazardous waste generated by 124,371 tons, and they conserved more than 5,000 acres of land (EPA 2007a). The results of the program, however, have not yet been extensively reviewed by outside researchers. As was true for 33/50, the key question about Performance Track remains whether the observed reductions by participating facilities might have occurred anyway under a business-as-usual scenario.

Environmental advocacy organizations criticized the Performance Track program, specifically after the Bush administration proposed offering Performance Track members additional modifications in regulatory requirements. The Natural Resources Defense Council (NRDC) objected to member incentives that would reduce monitoring, record-keeping, and reporting and shrink the level of EPA and state environmental agency oversight (Walke 2005). The Environmental Integrity Project (EIP), which advocates stronger enforcement of federal and state environmental laws, similarly called on EPA to step back from "ever more ambitious regulatory breaks" for Performance Track members (Ware 2006). Both NRDC and EIP raised concerns that Performance Track entry criteria failed to ensure that members demonstrated truly "superior" environmental performance that would warrant regulatory relief and other similar benefits. In a letter to then EPA administrator Stephen Johnson, EIP and 30 other environmental organizations called on the agency to delay expanding Performance Track until it could "show that the program's [societal] benefits justify reducing oversight, relaxing legal requirements, or excusing violations of the law" (Schaeffer 2006).

Despite such concerns, EPA in 2006 issued a further exemption allowing Performance Track plants with secondary containment facilities to self-inspect their hazardous waste tanks only once a month, in contrast to the daily or weekly requirements for non-Performance Track facilities (EPA 2006). This benefit was relatively modest, however, compared with those outlined in EPA's initial proposal.

EPA's own Office of the Inspector General, which is charged with evaluating agency programs, voiced apprehensions similar to those of the environmental groups. In a 2007 report, the inspector general found that some members' environmental performance fell below the average performance for their sector in terms of regulatory compliance and releases of toxic chemicals, and concluded that participation by those underperforming facilities undermined the integrity of the entire program (EPA 2007b).

Although EPA's Performance Track staff disputed the inspector general's findings, the program's die had been cast. Shortly after Barack Obama's election in 2008, the individual he would come to select as EPA administrator, Lisa Jackson, criticized Performance Track in the press, stating "I think it's just one of those window-dressing programs that has little value" (Sullivan and Shiffman 2008). It came as little surprise that one of the first decisions Jackson made as administrator was to shut down Performance Track.

Voluntary Programs in General

As both the 33/50 and NEPT programs illustrate, voluntary programs face major challenges in terms of determining whether they achieve credible results. Although it is possible for government to report that participants in voluntary programs achieve improvements in their performance, evaluating these programs presents obvious inferential obstacles. Would the participants in these programs have achieved these improvements anyway? Voluntary programs attract, after all, volunteers, who often act differently than non-volunteers. Rather than encouraging firms to make new environmental gains, voluntary programs could effectively be serving as a magnet to attract disproportionately those firms that already are performing better for other reasons (Kagan et al. 2002). According to a 2007 project that assessed the impact of seven voluntary energy and environmental programs in the United States, Europe, and Japan, all but one of the analyses suggested that voluntary programs affected behavior; focusing on energy-related activities, however, the effect was less than 10 percent and more typically closer to 5 percent (Morgenstern and Pizer 2007). None of the project's studies "found truly convincing evidence of dramatic environmental improvements," leading the authors to conclude, "We find it hard to argue for voluntary programs where there is a clear desire for major changes in behavior" (*184*).

NONGOVERNMENTAL VOLUNTARY PROGRAMS

Nongovernmental organizations also can establish voluntary programs. The most well-known and closely studied example has been the environmental management system standards issued by the International Organization for Standardization (ISO), the international standards-setting body. Established in 1947 in Geneva, Switzerland, ISO is made up of national standards bodies from more than 100 countries. Since its formation, ISO has developed more than 10,000 standards governing, among other things, paper sizes, the format of credit cards, and automobile dashboard symbols. ISO standards seek to ensure that products share uniform characteristics no matter where they are produced.

An environmental management system (EMS) is a regulatory structure that arises within an organization, a collection of internal efforts at policymaking, planning, and implementation that yields benefits for the organization as well as potentially significant benefits for society at large. Beginning in the early 1990s, national standards organizations in various countries began developing their own guidelines for how an EMS should be implemented. Around the same time, the European Commission also developed a standard for EMS implementation, known as the Eco-Management and Audit Scheme (EMAS). The plethora of EMS standards developed around the world led in 1996 to the adoption of the ISO 14001 standards.

Business leaders called on ISO to contribute toward sustainable business development as part of preparations for the United Nations Conference on Environment and Development, held in Brazil in 1992. These pressures, combined with the success of ISO 9000 standards on quality management, led ISO in 1992 to establish subcommittees to work on aspects of environmental management including EMSs, auditing, labeling, performance evaluation, and life-cycle assessment. The work of these committees resulted in the ISO 14000 series of standards. The first of these, ISO 14001, establishes EMS criteria. This is the only standard in the 14000 series to which a facility may be certified as having an EMS in compliance with ISO criteria.

Today ISO 14000 standards are the most widely recognized for EMS design throughout the world, and about 50,000 facilities are certified as having adopted an EMS that meets ISO standards. Most of these facilities are in Japan and Europe; only a few thousand facilities in the United States have been certified as meeting ISO standards (Prakash and Potoski 2006). Many companies reportedly implement certifiable EMSs to satisfy customer demands for required ISO certification of their suppliers. For example, both General Motors and Ford Motor Company in the past have required their parts suppliers to implement ISO-certified EMSs. At least for firms that supplied to GM or Ford under these purchasing requirements, then, ISO 14000 no longer was truly a *voluntary* program.

When ISO 14000 is truly voluntary, the key questions in assessing its effect on regulatory goals are similar to those in assessing governmental voluntary programs. Do firms that participate do better than they would have in the absence of the program? If participating firms perform better than nonparticipating firms, is this because nongovernmental voluntary programs such as ISO 14000 simply attract a disproportionate share of better-performing firms?

The relationship between ISO adoption and environmental performance has been widely studied, with varying results. A study of automotive facilities conducted prior to the decisions by Ford and GM to require the adoption of an EMS found that minimal reductions were the norm following ISO 14001 adoption in that sector (Matthews 2001). Similarly, Andrews et al. (2003) and Dahlstrom et al. (2003) could find no evidence of any systematic improvement in firms' performance. In contrast, a study of electronics firms that adopted ISO 14001 showed that these firms were generally able to catch up to best practices in their industry, especially if they began as a high producer of toxic emissions (Russo 2000). Anton et al. (2004) found similarly that ISO certification was associated with lower TRI emissions, and Dasgupta et al. (2000) linked it with better regulatory compliance. In the most systematic analysis to date, Prakash and Potoski (2006) found that ISO-certified facilities performed better than noncertified facilities in both emissions and regulatory compliance in terms of toxicity-weighted releases of TRI chemicals. Notably, though, they cautioned, "The scale of these improvements is not large, though it is statistically significant" (166). Even if ISO 14001 has had only minimal effects, it is always possible that other nongovernmental voluntary programs might fare better.

INDUSTRY SELF-REGULATION

Industry organizations also craft voluntary standards for their own members. Usually these organizations form around a sector, and they emerge when industries perceive a collective interest in protecting the reputation of their business. Particularly in dangerous industries, firms are "hostages to each other" (Rees 1996) in that a major accident occurring at one facility typically damages the reputations of all the rest—and prompts new government regulation of the entire sector. Indeed, industries have an incentive to engage in their own self-regulation in order to stave off future and potentially costlier government regulation (Lyon and Maxwell 2004). For reasons such as these, industry codes of conduct have proliferated over the past several decades (Nash and Ehrenfeld 1997).

By far the most well-known and extensively studied example of industry self-regulation has been the chemical industry's Responsible Care program, which grew out of a set of management principles adopted by the Canadian Chemical Producers Association in the late 1970s and early 1980s and gained international impetus after the Bhopal disaster in 1984. The media attention and public reaction to Bhopal was not lost on members of the chemical industry (Haufler 2001; Nash and Ehrenfeld 1997). In 1988, the main American chemical industry organization, the Chemical Manufacturers Association (CMA), adopted Responsible Care, calling on its members "to make health, safety, and environmental considerations a priority … for all existing and new products and processes" and "to develop and produce chemicals that can be manufactured, transported, used and disposed of safely" (CMA 1988). Through Responsible Care, CMA (now the American Chemistry Council) set standards for how firms were to manage their operations with respect to health, safety, and environment. For example, a pollution-prevention code aimed to encourage source reduction and recycling, while a product-stewardship code directed members to work with customers to ensure proper use and disposal of chemicals. From 1988 to 1994, CMA developed six codes in total (Nash and Ehrenfeld 1997). These codes, however, did not dictate how much pollution member firms emit or specific actions that should be taken to prevent worker accidents. Instead,

CMA expected its members to establish their own performance targets and identify means of meeting them.

Although adherence to the Responsible Care principles and codes was expected of all members, the trade association did not disclose information to the public about members' compliance nor did it remove firms from the association for failing to adhere to the codes. Instead, the association relied on informal controls. Members' progress was initially tracked by and known only to a consultant group hired by CMA. In 1996, the association took the further step of disclosing to the trade association's board the names of members that were failing to make adequate progress (Haufler 2001), and in 2000 it began ranking each member based on safety performance and circulating this ranking internally to the membership (Nash 2002). In 2002, the association undertook a major reform of Responsible Care, requiring each member by 2007 to obtain third-party certification of its environmental, health, and safety management system, as well as to begin disclosing to the public its environmental and safety record (Nash 2002).

Trade association codes of practice like Responsible Care are voluntary in the sense that firms do not have to adhere to them if they do not want to be members of the association. But some firms have significant incentives to belong to their sector's trade association, with benefits ranging from the exchange of research and development information to the collective interests of lobbying representation in Washington, D.C. For many firms, maintaining membership in their trade associations may be more important than meeting the criteria for entry into voluntary governmental or nongovernmental programs. Moreover, because trade association codes are created by industry itself, the demands they place on firms should be less costly. Thus we should expect to see greater participation in industry self-regulatory programs than in governmental voluntary programs.

It is less clear, though, the degree to which self-regulation leads firms to take their participation in these programs seriously and to adopt costly changes in practice. One study of Responsible Care showed a range of responses, with some firms taking compliance quite seriously, whereas others treated it more like a paperwork exercise (Howard et al. 1999). In another study, King and Lenox (2000) reported somewhat surprising results from a statistical analysis of Responsible Care. Controlling for facility size and using an index based on the toxicity of TRI releases, they found that CMA members reduced their releases more slowly than did comparable nonmember firms. Part of the explanation may have been a selection bias in the types of firms that join CMA. Indeed, the authors found that members tended disproportionately to be those firms that had higher reported TRI releases at the outset. They concluded, however, that this implies a weakness of CMA's self-regulation, because the organization did not succeed in bringing its members up to better performance levels. They also reasoned that Responsible Care contrasts with the successful account of self-regulation of the nuclear industry provided by Rees (1996) because that industry is smaller, less competitive, and works more closely with a single regulator (the Nuclear Regulatory Commission) that has an ability to sanction nuclear power plants that did not perform well. The chemical industry, by contrast, consists of a large number of firms, including a few large ones especially concerned about their image, but also many smaller firms that have less of a share in the collective reputation of the industry (Gunningham 1995). In short, the CMA faces a much more challenging collective action problem when it comes to self-regulation.

Trade association codes such as Responsible Care face inevitable questions of credibility and effectiveness. As already noted, the chemical industry has taken steps to strengthen the monitoring, disclosure, and sanctioning aspects of its program. Gunningham and Rees (1997) suggest that creating a successful industry system of self-regulation depends on identifying a clear set of moral standards and creating effective oversight institutions to provide accountability and transparency. In addition, they and others (e.g., Segerson and Miceli 1999) stress the importance of an external threat to convince member firms that it is in their long-term interest to cooperate and comply with the industry self-regulatory system. Without a credible external presence, as well as robust institutions for monitoring and sanctioning, the risk exists that self-regulatory institutions will serve as a smokescreen or a form of "greenwashing." Looking across a broad range of self-regulatory programs, Darnall and Carmin (2005) urge caution because "only a small portion take steps to ensure that participants reduce their environmental impacts."

IMPLICATIONS FOR NANOTECHNOLOGY

The fundamental challenge of any regulatory regime stems from a built-in conundrum: in order to advance broader policy goals, regulation seeks to induce businesses to act in ways they do not want to act all on their own. Just as James Madison said that government would not be needed if men were angels, regulation of some kind would not be needed if businesses already acted in socially optimal ways by internalizing their externalities or compensating for information asymmetries. But because the advantages of a market economy can be had only at the price of some market failures, regulation is essential—even if challenging.

Regulating well requires accurate information. To control externalities or ensure adequate product disclosure or safety, regulators need to know about the risks created by different types of products and production processes—both the nature and magnitude of any harmful activity or products, as well as the probability of harm occurring. Regulators also need to understand the causes of regulatory problems. Yet in contemplating the regulation of nanotechnology, this is the initial difficulty, and an overwhelming one. At present, the health and environmental risks associated with engineered nanomaterials are virtually unknown. This could be because such risks are nonexistent or, more likely, have yet to be identified.

In the absence of the necessary information about whether, whom, and how to regulate, it seems sensible for government to want to engage business by adopting flexible regulatory instruments, developing voluntary programs, or encouraging the development of nongovernmental codes or industry self-regulation. Surely these strategies would be preferable to any blind effort at conventional regulation imposed by government on a burgeoning industry with great potential for welfare-enhancing innovation. For these reasons, it is not surprising that governments and industry have so far addressed nanotechnology's potential health, safety, and environmental risks using strategies that fall within the categories described in this chapter.

To date, governments have avoided imposing highly prescriptive regulations on the manufacturing or use of nanomaterials. The only governmental entity in the United States that has so far imposed regulation specifically on nanotechnology—the city council of Berkeley, California—opted for an information disclosure approach. Adopted in 2006, the Berkeley ordinance requires "facilities that manufacture or use manufactured nanoparticles [to] submit a separate written disclosure of the current toxicology of the materials reported, to the extent known, and how the facility will safely handle, monitor, contain, dispose, track inventory, prevent releases and mitigate such materials" (City of Berkeley 2006).

Most governmental efforts have taken a decidedly voluntary approach. In 2007, the city of Cambridge, Massachusetts, considered whether to impose similar binding information disclosure requirements but opted instead for a strictly voluntary approach. The Cambridge Public Health Department, acting in conjunction with a citizen advisory committee, recommended a voluntary survey of local facilities that might be using nanomaterials and the creation of a technical assistance program to promote best practices in health and safety. The department based its decision to engage businesses in an "active and constructive collaboration" on an explicit recognition of the government's lack of information, noting the "great practical challenge to establish an evidence-based risk management framework for the safe production, manipulation, and disposal of engineered nanoscale materials given the large number of questions that remain unresolved" (CNAC 2008).

For much the same reason, EPA established in 2008 a voluntary Nanoscale Materials Stewardship Program (NMSP). Through NMSP, EPA seeks to encourage businesses that use, import, or manufacture nanomaterials to share "all known or reasonably ascertainable information regarding specific nanoscale materials," including information on "material characterization, hazard, use, potential exposures, and risk management practices" (EPA 2008). The agency believes that NMSP can help participants improve their environmental, health, and safety processes, as well as diffuse best practices more widely throughout industry. As of 2009, more than 30 companies that use or manufacture nearly 125 different nanomaterials had submitted information to EPA under the program, and 4 of these companies had committed to partner with the agency to engage in further research and data development on nanotechnology risks (EPA 2009). As with other EPA voluntary programs, the NMSP has not escaped criticism. The Environmental Defense Fund (EDF), partner with DuPont in a widely touted Nano Risk Framework (discussed below), charged that NMSP has generated only limited—even "skewed"—data, and that, even so, the agency has failed to release to the public the modest information it has been able to gather (EDF 2008).

A few nanotechnology-focused nongovernmental voluntary initiatives have also begun to appear. For example, ISO has established a technical committee to study nanotechnology and develop recommendations for the management of its risks. Among other things, the committee seeks to develop "risk assessment tools relevant to the field of nanotechnologies; … protocols for containment, trapping and destruction of nanoparticles and nanoscale entities; [and] occupational health protocols relevant to nanotechnologies" (ISO 2007). Eventually, nanotechnology may well be covered by new ISO management standards similar to ISO 14000.

The Brussels-based Nanotechnology Industries Association (NIA), formed in 2005, has already joined with private sector, public sector, and nongovernmental partners to establish a Responsible NanoCode. The NanoCode consists of seven highly broad principles, such as that nanotech firms "ensure high standards of occupational health and safety for its workers" and "carry out thorough risk assessments and minimize any potential public health, safety or environmental risks." A Responsible NanoCode working group has identified "examples of good practice" that are intended to illustrate how firms might act, consistent with the broad principles, although even these examples are broad. A "good practice" in environmental management includes any process "to identify, evaluate and minimize any risk to the general public, users or the environment" from nanotechnology (NIA et al. 2008). The Responsible NanoCode also emphasizes transparency and engagement with so-called stakeholders, including employees, customers, academics, shareholders, nongovernmental organizations, and the general public.

A notable self-regulatory initiative came out of the DuPont Corporation's partnership with the Environmental Defense Fund. After a series of dialog sessions and joint research, DuPont and EDF in 2007 issued a Nano Risk Framework that outlines a comprehensive risk management process for manufacturers and users of engineered nanomaterials. The framework is "designed to be flexible" but is explicitly "information-driven," in that it calls for gathering as much information as possible or for relying on "reasonable worst-case assumptions" (EDF–DuPont 2007). It also encourages transparency to the extent feasible without disclosing trade secrets or other confidential business information. DuPont has already applied the Nano Risk Framework to its development of three nanomaterials, and both DuPont and EDF hope that the framework will be used in the future by other companies working with nanomaterials.

These examples demonstrate that business plays the central role in society's current approach to protecting the public from any hazards associated with nanotechnology. This business-centered pathway parallels previous efforts to engage business in environmental protection, such as through flexible forms of regulation, voluntary government programs, nonbinding standards set by nongovernmental entities, and self-regulatory actions by industry (Eisner 2007).

It is much too early to know whether the current strategies for addressing nanotechnology risks will suffice to protect the public. But judging from previous efforts to engage and rely on business in regulatory governance, there appears to be some reason to hope that the current efforts in the nanotechnology realm can be at least somewhat effective—but also reason to fear that they will at best yield only modest outcomes. Engaging business in regulatory governance can be successful when firms use their flexibility within an overarching system of oversight by their industry peers, third parties, or government. Yet the very reasons that make delegating discretion to industry seem attractive—even necessary—in the context of nanotechnology are reasons to suspect the effectiveness of voluntary or discretionary efforts. The nanotechnology industry is neither small nor well organized. No clear overarching industry exists, so collective efforts akin to those of the nuclear industry, or even the chemical industry, will probably face much more significant obstacles. Smaller companies may be less likely, or less able, to engage in the extensive risk management protocols that larger firms support (Mandel 2008). Moreover, what governmental or third-party overseers should look for in this industry is still far from clear, as it is not even known whether engineered nanomaterials will prove as harmful as some have feared. Unfortunately, the same absence of information that makes government a poor central planner or regulator of nanotechnology also inhibits government's ability to act as an effective overseer of firms' own management plans or voluntary actions.

Uncertainty over nanotechnology's risks leaves society little choice but to rely more heavily on industry for information and protection than it does with respect to the risks from better-understood technologies. As such, engaging business in the task of regulating nanotechnology's risks will continue to be absolutely necessary. It is unclear, however, whether engaging business ultimately will be sufficient by itself to protect public health and the environment from any dangers that may loom.

REFERENCES

Ackerman, Bruce, and Richard Stewart. 1985. Reforming Environmental Law. *Stanford Law Review* 37: 1333–65.

Andrews, Richard N. L., Deborah Amaral, Nicole Darnall, Deborah Rigling Gallagher, Daniel Edwards, Jr., Andrew Hutson, Chiara D'Amore, Lin Sun, and Yihua Zhang. 2003. Environmental Management Systems: Do They Improve Performance? Project Final Report. National Database on Environmental Management Systems, University of North Carolina. ndems.cas.unc.edu/.

Anton, Wilma Rose Q., George Deltas, and Madhu Khanna. 2004. Incentives for Environmental Self-Regulation and Implications for Environmental Performance. *Journal of Environmental Economics and Management* 48 (1): 632–54.

Ayres, Ian, and John Braithwaite. 1992. *Responsive Regulation: Transcending the Deregulation Debate.* New York, NY: Oxford University Press.

Bardach, Eugene, and Robert A. Kagan. 1982. *Going by the Book: The Problem of Regulatory Unreasonableness.* Philadelphia, PA: Temple University Press.

Bennear, Lori Snyder. 2006. Evaluating Management-Based Regulation: A Valuable Tool in the Regulatory Toolbox? In *Leveraging the Private Sector: Management-Based Strategies for Improving Environmental Performance,* edited by Cary Coglianese and Jennifer Nash. Washington, DC: Resources for the Future, 51–86.

———. 2008. What Do We Really Know? The Effect of Reporting Thresholds on Inferences Using Environmental Right-to-Know Data. *Regulation & Governance* 2: 293–315.

Bi, Xiang, and Madhu Khanna, Impact of EPA's Voluntary 33/50 Program on Pollution Prevention Adoption and Toxic Releases, paper presented at the 2008 American Agricultural Economics Association Annual Meeting, Orlando, Florida. http://purl.umn.edu/6258.

Braithwaite, John. 1982. Enforced Self Regulation: A New Strategy for Corporate Crime Control. *Michigan Law Review* 80: 1466–507.

Breggin, Linda K., and Leslie Carothers. 2006. Governing Uncertainty: The Nano-technology Environmental, Health, and Safety Challenge. *Columbia Journal of Environmental Law* 31: 286.

Chinander, Karen R., Paul R. Kleindorfer, and Howard C. Kunreuther. 1998. Compliance Strategies and Regulatory Effectiveness of Performance-Based Regulation of Chemical Accident Risks. *Risk Analysis* 18: 135–44.

City of Berkeley. 2006. Municipal Code Section 15.12.040 on Manufactured Nanoparticle Health and Safety Disclosure. City of Berkeley, California. www.ci.berkeley.ca.us/citycouncil/2006citycouncil/packet/121206/2006-12-12%20Item%2003%20-%20Ord%20-%20Nanoparticles.pdf (accessed March 12, 2009).

CMA (Chemical Manufacturers Association). 1988. Responsible Care: Pollution Prevention Code. actrav.itcilo.org/actrav-english/telearn/global/ilo/code/responsi.htm (accessed March 11, 2009).

CNAC (Cambridge Nanomaterials Advisory Committee). 2008. *Recommendations for a Municipal Health & Safety Policy for Nanomaterials: A Report to the Cambridge City Manager.* Cambridge, MA: Cambridge Public Health Department.

Coglianese, Cary. 2007. Business Interests and Information in Environmental Rulemaking. In *Business and Environmental Policy,* edited by Michael Kraft and Sheldon Kamieniecki. Cambridge, MA: MIT Press, 185–210.

Coglianese, Cary, and David Lazer. 2003. Management-Based Regulation: Prescribing Private Management to Achieve Public Goals. *Law & Society Review* 37 (4): 691–730.

Coglianese, Cary, and Jennifer Nash. 2001. Environmental Management Systems and the New Policy Agenda. In *Regulating from the Inside: Can Environmental Management Systems Achieve Policy Goals?,* edited by Cary Coglianese and Jennifer Nash. Washington, DC: Resources for the Future, 1–26.

————. 2004. *The Massachusetts Toxic Use Reduction Act: Design and Implementation of a Management-Based Environmental Regulation*. Report RPP-07-2004. Cambridge, MA: Center for Business and Government, Kennedy School of Government.

Dahlstrom, Kristina, Chris Howes, Paul Leinster, and Jim Skea. 2003. Environmental Management Systems and Company Performance: Assessing the Case for Extending Risk-Based Regulation. *European Environment* 13 (4): 187–203.

Darnall, Nicole, and Joann Carmin. 2005. Greener and Cleaner? The Signaling Accuracy of U.S. Voluntary Environmental Programs. *Policy Sciences* 38: 71–90.

Dasgupta, Susmita, Hemamala Hettige, and David Wheeler. 2000. What Improves Environmental Performance? Evidence from Mexican Industry. *Journal of Environmental Economics and Management* 39 (1): 39–66.

Davies, J. Clarence. 2006. Managing the Effects of Nanotechnology. *PEN* 2. www.nanotechproject.org/process/assets/files/2708/30_pen2_mngeffects.pdf(accessed March 12, 2009).

————. 2007. EPA and Nanotechnology: Oversight for the 21st Century. *PEN* 9. www.nanotechproject.org/mint/pepper/tillkruess/downloads/tracker.php?url =http3A//www.nanotechproject.org/process/assets/files/2698/197_nanoepa_ pen9.pdf (accessed March 12, 2009).

EDF (Environmental Defense Fund). 2008. EPA Nanotechnology Voluntary Program Risks Becoming a 'Black Hole'. www.edf.org/pressrelease.cfm?contentID=8162 (accessed March 12, 2009).

EDF–DuPont (Environmental Defense Fund–DuPont). 2007. Nano Risk Framework. www.environmentaldefense.org/documents/6496_Nano%20Risk%20Frame work.pdf (accessed March 12, 2009).

Eisner, Marc Allen. 2007. *Governing the Environment: The Transformation of Environmental Regulation*. Boulder, CO: Lynne Rienner Publishers.

EPA (U.S. Environmental Protection Agency). 1999. 33/50 Program: The Final Record. EPA-745-R-99-004. www.epa.gov/oppt/3350/ (accessed March 12, 2009).

————. 2006. Resource Conservation and Recovery Act Burden Reduction Initiative. 71 Fed. Reg. 16,862, 16,881. April 4.

————. 2007a. Today's Commitments. Tomorrow's World. Five Years of Environmental Leadership. Performance Track Fifth Annual Progress Report. www.epa.gov/ perftrac/downloads/PTPRreport_05final.pdf (accessed March 12, 2009).

————. 2007b. Performance Track Could Improve Program Design and Management to Ensure Value. Report No. 2007-P-00013. www.epa.gov/oig/reports/2007/2007 0329-2007-P-00013.pdf (accessed March 12, 2009).

————. 2008. Nanoscale Materials Stewardship Program. 73 Fed. Reg. 4,861, 4,862-63. January 28.

————. 2009. Nanoscale Materials Stewardship Program. www.epa.gov/oppt/nano/ stewardship.htm (accessed March 12, 2009).

Ferrell, Allen. 2007. Mandated Disclosure and Stock Returns: Evidence from the Over-the-Counter Market. *Journal of Legal Studies* 36 (2): 213–51.

Fung, Archon, Mary Graham, and David Weil. 2007. *Full Disclosure: The Perils and Promise of Transparency.* Cambridge, UK: Cambridge University Press.

Fung, Archon, and Dara O'Rourke. 2000. Reinventing Environmental Regulation from the Grassroots Up: Explaining and Expanding the Success of the Toxics Release Inventory. *Environmental Management* 25 (2): 115–27.

Gamper-Rabindran, Shanti. 2006. Did the EPA's Voluntary Industrial Toxics Program Reduce Emissions? A GIS Analysis of Distributional Impacts and By-Media Analysis of Substitution. *Journal of Environmental Economics and Management* 52: 391–410.

Graham, Mary. 2002. *Democracy by Disclosure: The Rise of Technopopulism.* Washington, DC: Brookings Institution.

Gunningham, Neil. 1995. Environment, Self-Regulation, and the Chemical Industry: Assessing Responsible Care. *Law & Policy* 17 (1): 57.

———. 1996. From Compliance to Best Practice in OHS: The Role of Specification, Performance, and Systems-Based Standards. *Australian Journal of Labor Law* 9: 221–46.

Gunningham, Neil, and Richard Johnstone. 1999. *Regulating Workplace Safety: Systems and Sanctions.* Oxford, UK: Oxford University Press.

Gunningham, Neil, and Joseph Rees. 1997. Industry Self-Regulation: An Institutional Perspective. *Law & Policy* 19: 363–414.

Hahn, Robert W. 1989. *A Primer on Environmental Policy Design.* New York, NY: Harwood Academic Publishers.

Hahn, Robert W., and Gordon L. Hester. 1989. Marketable Permits: Lessons for Theory and Practice. *Ecology Law Quarterly* 16: 361–406.

Hamilton, James T. 2005. *Regulation through Revelation: The Origin, Politics, and Impacts of the Toxics Release Inventory Program.* Cambridge, UK: Cambridge University Press.

Haufler, Virginia. 2001. *A Public Role for the Private Sector: Industry Self-Regulation in a Global Economy.* Washington, DC: Carnegie Endowment for International Peace.

Howard, Jennifer, Jennifer Nash, and John Ehrenfeld. 1999. Industry Codes as Agents of Change: Responsible Care Adoption by U.S. Chemical Companies. *Business Strategy and the Environment* 8 (5): 281–95.

Hutter, Bridget. 2001. *Regulation and Risk: Occupational Health and Safety on the Railways.* Oxford, UK: Oxford University Press.

Innes, Robert, and Abdoul G. Sam, 2008. Voluntary Pollution Reductions and the Enforcement of Environmental Law: An Empirical Study of the 33/50 Program. *Journal of Law & Economics* 51 (2): 271–96.

ISO (International Organization for Standardization). 2007. Business Plan ISO/TC 229 Nanotechnologies. isotc.iso.org/livelink/livelink/6356960/TC_229_BP_2007-2008.pdf?func=doc.Fetch&nodeid=6356960 (accessed March 11, 2009).

Jin, Ginger Zhe, and Philip Leslie. 2003. The Effect of Information on Product Quality: Evidence from Restaurant Hygiene Grade Cards. *Quarterly Journal of Economics* 118 (2): 409–51.

Kagan, Robert A., Neil Gunningham, and Dorothy Thornton. 2002. Explaining Corporate Environmental Performance: How Does Regulation Matter? *Law & Society Review* 37: 51–90.

Karkkainen, Bradley. 2001. Information as Environmental Regulation: TRI, Performance Benchmarking, Precursors to a New Paradigm? *Georgetown Law Journal* 89: 257–370.

Khanna, Madhu. 2006. The U.S. 33/50 Voluntary Program: Its Design and Effectiveness. In *Reality Check: The Nature and Performance of Voluntary Environmental Programs in the United States, Europe, and Japan,* edited by Richard D. Morgenstern and William A. Pizer. Washington, DC: Resources for the Future, 15–42.

Khanna, Madhu, and Lisa A. Damon. 1999. EPA's Voluntary 33/50 Program: Impact on Toxic Releases and Economic Performance of Firms. *Journal of Economics and Management* 37 (1): 1–25.

King, Andrew A., and Michael Lenox. 2000. Industry Self-Regulation without Sanctions: The Chemical Industry's Responsible Care Program. *Academy of Management Journal* 43 (4): 698–716.

Kleindorfer, Paul R. 1999. Understanding Individuals' Environmental Decisions: A Decision Science Approach. In *Better Environmental Decisions: Strategies for Governments, Businesses, and Communities,* edited by Ken Sexton. Washington, DC: Island Press, 37–56.

———. 2006. The Risk Management Program Rule and Management-Based Regulation. In *Leveraging the Private Sector: Management-Based Strategies for Improving Environmental Performance,* edited by Cary Coglianese and Jennifer Nash. Washington, DC: Resources for the Future, 87–110.

Kleindorfer, Paul R., and Eric Orts. 1998. Informational Regulation of Environmental Risks. *Risk Analysis* 18: 155–70.

Lyon, Thomas P., and John W. Maxwell. 2004. *Corporate Environmentalism and Public Policy.* Cambridge, UK: Cambridge University Press.

Mandel, Gregory. 2008. Nanotechnology Governance. *Alabama Law Review* 59: 1323–84.

Marsh & McLennan. 1997. *Large Property Damage Losses in the Hydrocarbon-Chemical Industries: A Thirty-Year Review.* 17th ed. New York, NY: Marsh & McLennan.

Matthews, Deanna H. 2001. Assessment and Design of Industrial Environmental Management Systems. PhD diss, Carnegie Mellon University.

Morgenstern, Richard D., and William A. Pizer. 2007. *Reality Check: The Nature and Performance of Voluntary Environmental Programs in the United States, Europe, and Japan.* Washington, DC: Resources for the Future.

Nash, Jennifer. 2002. Industry Codes of Practice: Emergence and Evolution. In *New Tools for Environmental Protection: Education, Information, and Voluntary*

Measures, edited by Thomas Dietz and Paul C. Stern. Washington, DC: National Academy Press, 235–52.

Nash, Jennifer, and John Ehrenfeld. 1997. Codes of Environmental Management Practice: Assessing Their Potential as a Tool for Change. *Annual Review of Energy and the Environment* 22: 487–535.

NIA (Nanotechnology Industries Association), Insight Investment, the Royal Society, and the Nanotechnology Knowledge Transfer Network. 2008. The Responsible Nano Code. www.responsiblenanocode.org/documents/TheResponsibleNanoCo deUpdateAnnouncement.pdf (accessed March 12, 2009).

OMB (Office of Management and Budget, Office of Information and Regulatory Affairs). 2000. Guidelines to Standardize Measures of Costs and Benefits and the Format of Accounting Statements. www.whitehouse.gov/omb/memoranda/m00-08.pdf (accessed March 12, 2009).

Pildes, Richard H., and Cass R. Sunstein. 1995. Reinventing the Regulatory State. *University of Chicago Law Review* 62: 1–129.

Prakash, Aseem, and Matthew Potoski. 2006. *The Voluntary Environmentalists: Green Clubs, ISO 14001, and Voluntary Environmental Regulations.* Cambridge, UK: Cambridge University Press.

Rees, Joseph V. 1996. *Hostages of Each Other: The Transformation of Nuclear Safety since Three Mile Island.* Chicago, IL: University of Chicago Press.

Richards, Kenneth. 2000. Framing Environmental Policy Choice. *Duke Environmental Law & Policy Forum* 10: 221–85.

Russo, Michael V. 2000. Institutional Change and Theories of Organizational Strategy: ISO 14001 and Toxic Emissions in the Electronics Industry. Paper presented at the 60th Annual Meeting of the Academy of Management. Toronto, Ontario, Canada.

Sam, Abdoul G., and Robert Innes. 2004. Voluntary Pollution Reductions and the Enforcement of Environmental Law: An Empirical Study of the 33/50 Program. Research Paper 2004-08. Cardon Research Papers in Agricultural and Resource Economics, University of Arizona.

Schaeffer, Eric. 2006. Letter to the Honorable Stephen Johnson, Administrator, U.S. Environmental Protection Agency, January 25, 2006. On file with the author.

Segerson, Kathleen, and Thomas J. Miceli. 1999. Voluntary Approaches to Environmental Protection: The Role of Legislative Threats. In *Voluntary Approaches in Environmental Policy*, edited by Carlo Carraro and Francois Láveque. Dordrecht, Netherlands: Kluwer Academic Publishers, 105–120.

Sinclair, Darren. 1997. Self-Regulation versus Command and Control? Beyond False Dichotomies. *Law & Policy* 19: 529–59.

Stavins, Robert N. 1989. Harnessing Market Forces to Protect the Environment. *Environment* 31: 28–35.

———. 2003. Experience with Market-Based Environmental Policy Instruments. In *Handbook of Environmental Economics*, 1, edited by Karl-Goran Maier and Jeffrey Vincent. Amsterdam, Netherlands: Elsevier Science, 355–435.

Stokey, Edith, and Richard Zeckhauser. 1978. *A Primer for Policy Analysis.* New York, NY: W. W. Norton & Co.

Sullivan, John, and John Shiffman, 2008. Green Club an EPA Charade, *Philadelphia Inquirer,* A1 (December 9).

Sunstein, Cass R. 1999. Informational Regulation and Informational Standing: Akins and Beyond. *University of Pennsylvania Law Review* 147: 613–75.

Sylvester, Douglas J., Kenneth W. Abbott, and Gary E. Marchant. 2008. Risk Management Principles for Nanotechnology. *NanoEthics* 2: 43–60.

Tietenberg, Thomas. 1990. Economic Instruments for Environmental Regulation. *Oxford Review of Economic Policy* 6: 17–33.

Vidovic, Martina and Neha Khanna, 2007. Can Voluntary Pollution Prevention Programs Keep Their Promises? *Journal of Environmental Economics and Management,* 53 (2): 180–95.

Viscusi, W. Kip. 1983. *Risk by Choice: Regulating Health and Safety in the Workplace.* Cambridge, MA: Harvard University Press.

Walke, John. 2005. Letter to EPA Docket ID OA-2005-0003 on behalf of the Natural Resources Defense Council. November 3.

Ware, Patricia. 2006. Benefits of "Performance Track" Program in Question, Environmental Group Says. *Environment Reporter* 37 (6): 309–10.

CHAPTER 5

EPA AND NANOTECHNOLOGY: THE NEED FOR A GRAND BARGAIN?

Marc Landy

This chapter has two distinct parts. The first describes the current institutional capacity of the U.S. Environmental Protection Agency (EPA) to regulate nanotechnology and explores alternative regulatory strategies. The second presents a political argument about why current conditions exist. My argument is that only by understanding the political dynamics that have shaped the agency is it possible to assess the political feasibility of any proposals for improving it. I conclude by examining the possibility of devising a political "grand bargain" between industry and environmentalists that would enable EPA to improve its capability to regulate nanotechnology.

EPA INCAPACITY AND NANOTECHNOLOGY REGULATION

To pursue a complex regulatory goal, an agency must have sufficient resources and the right mix of expertise. But even having the richest pool of resources and best available and appropriately skilled talent will not enable an agency to provide the right answers if it does not—or cannot—ask the right analytical questions. In our 1990 study on EPA, my coauthors and I maintained that the questions that need to be asked about potential harm to the environment have to be "educationally and strategically provocative":

> They focus agency and public attention on the choices that are available and the ethical issues those choices raise … They must reveal the moral and scientific presuppositions that underlie designations such as "carcinogen" and "hazardous substance." They must acknowledge that nature's inherent unpredictability renders outcomes uncertain. And, they must confront the limitations that the available scientific information places upon the ability to take effective action. (Landy et al. 1990, *320–21*)

Failing to inquire into the moral and scientific presuppositions underlying these most critical definitional issues threatens to set the agency adrift. Lacking the independent intellectual underpinnings for setting priorities and making choices, the agency cannot develop the political and institutional wherewithal to do either. Instead, priorities are set by the most powerful external forces that act on it, be they the courts, mass media, the regulated industry (especially under the Republicans), or environmental groups (especially under the Democrats).

Over the past three decades, the major regulatory instrument EPA has had at its disposal for addressing risks to human health and the environment created by chemical substances is the Toxic Substances Control Act (TSCA). Therefore, the next section examines TSCA and EPA's implementation of the act to determine whether the agency has asked good questions about how to address the risks chemical substances pose and, looking forward, whether it is well poised to ask good questions with regard to nanomaterials. In contrast to Marc Eisner's focus in Chapter 3 on the underlying statutory relevance of TSCA to nanomaterials, here I look at EPA as the law's implementing agent, concluding that, as currently conceived, EPA's implementation of TSCA renders the agency unable to obtain the information necessary to arrive at intellectually and scientifically defensible risk assessment. Moreover, the way the agency asks questions about risk is inadequate, rendering it unable to reach sound conclusions even if did have sufficient data at its disposal.

Thinking about Risk

The hardest and most important issue involving chemical regulation regards how much of a risk a particular chemical poses to human health or elements of the natural environment. Because available evidence is always insufficient and subject to alternative interpretation, the determination of risk will depend on a prior determination of what evaluatory principle to invoke in the face of this empirical confusion and uncertainty. Thus the most fundamental question the agency has to ask is which overarching principle to employ. Many in the environmental community would argue that because the public health (versus ecological) aim of environmental protection policy is to keep human beings safe, any ambiguous evidence regarding risk to human health should be interpreted in the most conservative plausible light. That is, if reputable scientific evidence indicates a chemical's risk, then that chemical must be regulated in some way. In this view, a chemical is "guilty until proven innocent." It cannot be exonerated by the mere existence of contrary evidence, unless that evidence is actually capable of denying the validity of the damning evidence.

An alternative approach finds a receptive audience, especially within the regulated community. In this view, regulators should emulate criminal law standards and treat chemicals as "innocent until proven guilty." If the scientific evidence regarding the human health effects of a particular chemical is ambiguous, as it almost always is, then the interests of society are best served by allowing the chemical in question to continue to be produced and used, enabling society to continue to reap its benefits at least until a relatively unambiguous finding of harm emerges. In this view, evidence of risk should not be considered dispositive unless it is overwhelmingly more credible and ample than the contrary evidence.

These starkly opposing points of view are far more cogent and convincing in their rebuttals of the other's argument than in defense of their own. In fact, it is rarely possible to prove either the guilt *or* innocence of a chemical. Therefore, presuming its guilt on the basis of *some* evidence risks depriving society of great potential benefits that might come from putting or keeping the chemical in production. If the substance is used in medicine or pollution mitigation, such a conservative regulatory posture might even produce a net increase in human health risk or environmental degradation. On the other hand, waiting for conclusive proof that a chemical poses dangers exposes humans and the environment to potentially grave risks in what is likely to prove to be a very long, perhaps interminable, interval of time.

Experience obtained from decades of debates—and lawsuits—over risks posed by chemicals suggests that a better principle for assessing risk questions abandons the absolutism of the two principles described above and rests instead on comparative risk. Regulatory schemes that rely on the Sisyphean search for unambiguous guilt or innocence will unduly delay the introduction of potentially valuable new chemicals, if we must wait until they are proven innocent, or delay the imposition of potentially lifesaving regulatory controls, if we must wait until they are proven guilty. Further, a focus on comparative risk acknowledges the centrality of opportunity cost considerations and, therefore, of the adoption of a comparative risk framework (Davies 1996). Comparative risk calculation requires the consideration of not only the risks reduced by controlling the production or use of a certain chemical, but also any heightened risks incurred by its decreased availability as well as the use of any chemicals that serve as substitute for it. When we think in comparative risk terms, the limitations of either the "innocent until proven guilty" approach or its converse become clearer. It can no longer be said in defense of the former, "One can never be too safe," because it might well be the case that exaggerated caution with regard to one substance invites insufficient caution with regard to the overall health or environmental risks that actually exist. But nor can it simply be contended that a chemical is beneficial without comparing its potential risks and benefits to possible substitutes.

The proposed alternative approach has been termed "weight of evidence" (Rhomberg 2007). As the term implies, this approach tries to distinguish weak from strong evidence for harm, but it does not require an unambiguous finding of harm. It establishes criteria for sifting through the evidence to establish the relative plausibility and cogency of differing findings in order to determine, on balance, whether the evidence for a finding of harmful health or environmental effects is stronger or weaker than the case against. Not only does this provide a rationale for at least provisionally determining whether a substance should be viewed as worthy of regulatory control, but it also allows for making comparisons to potential substitutes based on the relative strength of the weight of evidence in regard to the risk these alternatives pose.

Because it asks far more nuanced and difficult questions, however, the weight of evidence approach places far greater strain on agency organizational capacity. It requires complex and subtle weighing of conflicting evidence, the devising of appropriate weighting criteria, and the creation of metrics for comparing the relative weights of evidence among potential substitutes. No single agency official is sufficiently knowledgeable and wise to arrive at such judgments on his or her own. Rather, assessing the weight of evidence is a collective task involving people with diverse forms of expertise and professional competency. Ensuring not only that such people exist in the agency, but also that they have the freedom and incentive to engage in such deliberations places strains on the personnel capacity of the agency and the flexibility of its institutional design. This requires the employment of personnel intellectually capable of engaging in such difficult endeavors, as well as the cultivation of relationships among personnel with different views so that a thoughtful deliberation can occur among them with regard to this complex task. Further, it necessitates a transparent public education approach capable of explaining how and why the agency has decided to either regulate or ignore a particular chemical despite the existence of contradictory evidence.

TSCA as Regulatory Framework

The Toxic Substances Control Act (TSCA) of 1976 (see U.S. Congress 1976) authorizes EPA to examine both new and existing chemicals in order to determine whether the uses of those chemicals pose unreasonable risks to human health or the environment. Where "unreasonable risk" is found, EPA is granted the statutory authority to regulate the importation, manufacture, distribution, use, and disposal of the chemical in question (Fletcher et al. 2008). The tools that the agency can deploy to address the risks such a chemical poses range in degrees of stringency: EPA can prohibit its production and distribution entirely; restrict the uses for which it can be produced or distributed; limit the volume or concentration in which it is produced; specify its means of disposal; or mandate that a warning label be placed on its container. Recognizing that the obligatory premarket screening cannot be considered definitive regarding present and future risks, the agency may require producers to continue to maintain and report records regarding the chemical's molecular structure; the uses to which it is put; the volume of its manufacture, use, and disposal; its environmental and health effects; and the number of people exposed to it in the course of its production, distribution, use, and disposal.

New Chemicals

TSCA's regulatory scheme is very different concerning new versus existing chemicals. A "new chemical" is defined as one that is not already on the TSCA Inventory. Of the roughly 83,000 substances currently on the inventory, 61,000 were on the original list that EPA published in 1979. The others were initially reviewed as new chemicals and then added to the inventory (GAO 1994; Phillips 2006; Denison 2009).

Manufacturers, importers, and processors planning to introduce a new chemical or produce, process, or use an existing chemical in a new way must notify EPA 90 days in advance by submitting a premanufacture notice (PMN). The PMN contains information on the chemical's identity, physical characteristics, processing and use, and available toxicity data. The agency then determines whether the substance poses an unreasonable risk. If so, the EPA administrator must issue regulations to sufficiently diminish that risk. The party submitting the PMN must include with it any test data on risks to human health or safety if available, but in practice, only about 15 percent of premanufacture notices include such data. In short, because PMN requirements do not compel manufacturers to do expensive and time-consuming testing, for the most part they do not do it (GAO 2005). If the PMN submitter does not hear from EPA within 90 days of PMN submission, it may begin manufacture or import of the substance after submission of a Notice of Commencement of Manufacture or Import (NOC). Submission of the NOC adds the substance to the inventory and changes its status from a new to an existing chemical.

The review process can take longer—sometimes considerably longer—than 90 days. The standard against which EPA assesses the "registerability" of a substance is whether it poses unreasonable risks of injury to health or the environment. If EPA believes a new substance poses such risk, then the submitter has every reason to grant the agency more time to conclude otherwise and will typically acquiesce to EPA's request that it "toll" the 90-day clock. Absent tolling, the submitter must endure a finding that the substance may pose unreasonable risks or withdraw the PMN to avert this finding and the adverse implications it inspires.

If EPA flags concern with the substance, then PMN resolution can take several forms. EPA may resolve its initial concerns about risk and drop the PMN from further review, clearing the way for commercial manufacture. It may require the development and submission of toxicity or other data as a condition of approval—which is costly and time-consuming, with uncertain results. Or the agency may impose limitations on the substance's manufacture, use, distribution, or disposal—terms that could make it less competitive in the marketplace. Finally, EPA could conclude that the substance does not meet the TSCA standard for registration and take steps to prohibit its manufacture. Although the decision to cancel may be appealed, the process is costly and the result uncertain, and commercial manufacture is disallowed during the proceeding (Bergeson 2007).

Lacking sufficient and conclusive data regarding the new chemical's properties, EPA typically relies on analogies. It conducts a structure activity relationships analysis (SAR) to screen and evaluate a chemical's toxicity. The

SAR relies on models comparing the new chemicals to those with similar molecular structures for which more compelling test data on health and environmental effects are available. These models are good at predicting some chemical characteristics, but not others. A recent joint EPA and EU project compared predictions of individual physical and chemical properties and health and environmental effects generated by SAR-based models with actual test data submitted to the EU. EPA's predictions varied depending on the effect or property subject to comparison. For example, EPA's models were capable of identifying substances that degraded slowly, but not those that degraded quickly (Bergeson 2007). They were correct only 57 percent of the time when predicting systemic toxicity and also consistently tended to underestimate health effects (Davies 2007).

In the three decades since the passage of TSCA, EPA has reviewed 32,000 new chemicals and required the manufacturers to reduce the risks associated with more than 3,500 of them. It has ordered the makers of 1,200 new chemicals to reduce exposure of production workers or perform toxicity tests when production exceeds a specified volume; for roughly 570 of them, manufacturers were required to submit premanufacture notices for any significant new uses of the chemical, enabling EPA to reevaluate risk levels caused by the different exposure circumstances these new uses would create. In 1,600 cases, manufacturers withdrew a chemical, often after receiving warning from the agency that it planned to prohibit or limit production or impose other forms of control (GAO 1994).

Existing Chemicals

By contrast, EPA is enabled to regulate an existing chemical listed on the TSCA Inventory only if it has "a reasonable basis" to conclude that the substance presents an "unreasonable risk" of injury to human health or the environment. The agency's ability to make such a determination is severely hampered by the difficulty it encounters in obtaining the necessary information from the manufacturer. The maker of an existing chemical need provide EPA with toxicity data only if the agency first determines that the chemical warrants such testing and promulgates a special rule mandating that the manufacturer do so. But where are such data to come from, if not from the manufacturer? Therefore, EPA lacks sufficient data to even decide whether to ask for more data. As a result, it has not been very active in scrutinizing existing chemicals, which continue to make up the bulk of the TSCA Inventory. It has performed its own internal reviews on only 2 percent of 83,000 eligible chemicals and required testing of fewer than 200

(Denison 2009). Of those, it has chosen to regulate only 5 chemicals—polychlorinated biphenyls (PCBs), fully halogenated chlorofluoroalkanes, dioxin, asbestos, and hexavalent chromium—in several instances many decades after widespread use or exposure began.

If EPA finds that an existing chemical does require regulation, the statutory authority granted to it by the TSCA requires that the agency impose the "least burdensome requirements" necessary to deal with that risk. And in this regard, the agency does not necessarily enjoy the discretion to decide what is "least burdensome" to the regulated industry. For example, the agency's most ambitious effort to control an existing chemical, the ban it imposed on the manufacture and use of asbestos, was overturned by the U.S. Fifth Circuit Court of Appeals because the court found that the agency had not indeed imposed the "least burdensome" control requirements (*Corrosion Proof Fittings v. EPA* 1991). The gist of the court's finding was that in deciding to ban the manufacture and sale of asbestos, EPA had failed to evaluate the costs and benefits of less onerous possible control options. Therefore, the agency was in no position to claim that a very burdensome regulation—an outright ban—was the least burdensome alternative. Nor had the agency produced evidence on the risks posed by the most likely substitutes to be employed for various purposes currently served by products containing asbestos. The court allowed the ban to remain in force for a few specific products that it viewed as especially dangerous—flooring felt, rollboard, and corrugated, commercial, or specialty paper—and upheld the ban on new uses of asbestos. But it lifted the ban on all other existing uses, including the many different asbestos products used in the construction and automobile industries.

Should Nanomaterials Be Considered New or Existing?

As the foregoing discussion indicates, EPA has had little success in regulating existing chemicals through TSCA. Although great informational obstacles exist for regulating new chemicals, the burdens that TSCA, and the judiciary's interpretation of the law, place on regulating chemicals already on the inventory are in fact far greater. Thus it is not surprising that those seeking to reduce the potential regulatory burden on nanomaterials, at least on those that are nanoversions of existing chemicals, claim that they should be treated as existing chemicals for purposes of TSCA regulation, whereas those seeking more stringent regulation insist that they should be treated as new chemicals (DeLisi 2009; Wright 2009).

In July 2007, EPA issued a position paper stating unambiguously that nanoversions of existing chemicals were, for purposes of TSCA, existing chemicals. This position was staked on its view that "nano" refers merely to physical size, not molecular structure. The agency acknowledged, however, that differences in physical size might imply differences in physical or chemical properties. For example, nanoscale versions of materials on the Chemical Substance Inventory, such as silver or titanium, may have very different physical characteristics and properties than do their more conventional forms, and therefore their nanoscale versions may well pose different types and levels of risk. But the agency maintained that it could not take those differences into account, because TSCA's definition of an existing chemical was based exclusively on "molecular identity." Whatever other differences, nanoscale versions of existing chemicals have the same molecular identity as their nonnano counterparts (EPA 2007). Nanomaterials that do not have counterparts on the inventory will be treated as new chemicals, but this treatment in no way differentiates them from any other type of new chemical.

Environmental advocates point out that EPA's reading of TSCA is not the only possible one. The agency has, albeit rarely and under very special circumstances, taken physical attributes as well as molecular structure into account when making a new chemical determination (EDF 2006). But EPA's interpretation, based as it is on a literal reading of the relevant statutory wording, and in line with its past practice, would seem to be neither arbitrary nor capricious and is therefore likely to survive judicial scrutiny.

Significant New Use

The American Bar Association has pointed out that TSCA does provide an alternative means for evading the cumbersome regulatory requirements of the existing chemicals regimen by declaring a nanoversion to be a "significant new use," and that Section 5(a)(2) of TSCA authorizes the agency to promulgate significant new use rules (SNURs) when an existing chemical is put to new purposes. Once a SNUR is issued, EPA can regulate the chemical as if it were a new chemical and subject it to the PMN process (ABA 2006).

EPA has issued only 41 SNURs for existing chemicals in 30 years of TSCA's existence, however. The vast bulk of SNURs have been issued for new chemicals as an adjunct of the PMN process. This may be because in order to issue a SNUR under Section 5(a)(2), EPA must promulgate a rule subject to public notice and comment, whereas under Section 5(a)(1), the agency already has in place a generic rule requiring submission of a notice. In any event, the SNUR is available when the nanoversion of a substance constitutes a use that is different from that of the conventional substance, which is likely to be the case much of the time. Whether EPA will avail itself of that option is an open question, given information scarcity and potential for litigation about what constitutes "least burdensome" regulatory actions.

EPA AND RISK ASSESSMENT

Even if obstacles created by TSCA to obtaining the necessary information were removed, EPA as an organization is not currently capable of asking the appropriate questions about risk. As discussed at the beginning of this chapter, the issue of risk assessment is highly politicized. Disputes over how to conduct risk assessment mask fundamental disputes about how to interpret ignorance and uncertainty. To restate the fundamental question: are suspect chemicals innocent until proven guilty, or the reverse?

EPA's risk assessment difficulties can best be illustrated by examining its efforts to deal with the most fully studied and best understood aspect of human health risk from toxic chemicals—cancer. The agency issued one set of Guidelines for Carcinogen Risk Assessment in 1986 and a revised set in 2005 (EPA 2005). The 1986 version epitomized the "guilty until proven innocent" approach. For practical purposes with respect to many of the substances being considered, if any two of the animal studies being reviewed showed tumors, the substance was to be deemed a "probable human carcinogen." This designation was made even if the other animal studies and human epidemiological evidence were negative or inconclusive or the methodology employed in the positive studies was less credible than in the other studies.

In response to criticism from both industry and risk assessment professionals, the 2005 guidelines sought to introduce greater flexibility regarding carcinogenicity assessment. They provided a wide-ranging account of the various statistical and inferential issues at stake in trying to evaluate conflicting evidence from the different types of animal and human epidemiological studies available regarding a particular suspect carcinogen. The guidelines recommended arriving at a weight of evidence determination based on such an evaluation (EPA 2005). But they did not actually provide either a conceptual framework or a metric by which to weigh diverse and contradictory outcomes. The 1986 guidelines, although rigid, at least provided such a metric: "two positives and the substance is guilty." The greater flexibility of the 2005 guidelines was not accompanied by any clear principle for making such a determination. They established the universe of issues that needed to be considered, but in the absence of standards for considering trade-offs between conflicting findings, they provided insufficient grounds for determining in which direction the weight of evidence points.

The revised guidelines were also insufficiently comprehensive in prescribing the types of evidence to be considered. They stressed the role of "positive" evidence—finding particular tumors in epidemiological studies or animal bioassays—and placed much less emphasis on the role of "negative" evidence, the lack of the same effects in other human studies or animal tests. This imbalance can lead to what can be seen as irrational conclusions. For example, if extensive testing of a substance produces a number of positive responses, but replications of those studies come up negative, the "weight of evidence" criteria in the guidelines would lead to the conclusion that the evidence for human carcinogenicity was strong. The alternative and more plausible interpretation, however, is that an inability to replicate those findings more likely indicates that the sporadic positive findings were false positives and therefore do not establish a strong case for human risk (Rhomberg 2007).

The Need for Hypothesis Testing

The lack of explicit discussion of trade-offs among types of evidence and between positive and negative findings is symptomatic of the agency's deeper problem: the inability to specify the underlying logic on the basis of which it might choose to give greater weight to certain findings and categories of findings and not to others. The basis for such logic has to be some hypothesis about how and why a particular substance does or does not cause cancer. In the absence of such a hypothesis, no rationale exists for deducing that results found in the laboratory will actually cause illness to exposed humans. The great virtue of a causal hypothesis is that it can be shown to be wrong or, at a minimum, to require other more or less plausible subsidiary hypotheses to justify its inability to explain the facts on its own. In the absence of such a hypothesis, one is left, in most cases, with nothing but a "gut feeling" for why the positive findings should be given more weight than the negative ones, or vice versa.

Only a clearly stated hypothesis about cause and effect is capable of turning data into evidence. Data acquire evidentiary importance only when they either do or do not substantiate a particular explanation of the phenomenon under investigation. Contrary to the "innocent until proven guilty" or "guilty until proven innocent" approaches, the hypothesis-driven approach does not claim to prove anything. Rather, it enables one to judge whether the available data more strongly supports the hypothesis being put forward or provides evidence for doubting it (Rhomberg 2007). The imperative to state a causal hypothesis is as important for assessing risks posed by nanomaterials as it is for any other form of environmental risk.

Having now established a rationale for explaining the evidence, one can then judge where the weight of the evidence lies. The weight is heavier if the hypothesis is consistent with the results of many studies and inconsistent with few, and the predictions it makes receive experimental confirmation. It receives additional weight if it requires few, if any, additional assumptions in order to remain consistent with the evidence, and those few assumptions are indeed plausible. The weight of evidence lightens if results across various sexes, species, strains, and doses do not accord with the predictions of the hypothesis. It lightens further the more it relies on additional, unproven, ad hoc assumptions to explain away empirical inconsistencies. Those assumptions may eventually turn out to be valid, but the hypothesis at stake is weakened to the extent that they have to be made while there is as yet no evidence to support them. Once the weight of evidence has been established for a particular hypothesis, it can then be compared with the weight of evidence for alternatives, including one that claims no risk to humans exists (Rhomberg 2007).

The hypothesis testing/weight-of-evidence approach is far from foolproof, because it can be no better than the evidence at hand permits. Its greatest virtue is that it reveals the logic that lies behind the conclusion reached and therefore enables a deliberation to take place in which the strength of that logic is compared with the logic behind alternative modes of explanation. Such a deliberation will never be free of partisanship. Those with differing ideological or economic agendas may still differ about matters of interpretation, but at least it provides a transparent basis for discussion. Rather than simply defend their preferred adversarial standard—guilty until proven innocent or the reverse—the contending parties must explain why they do or do not agree with weight-of-evidence judgments reached by the agency. Often the agency will be in a position to point out the illogic of such challenges or propose further empirical studies or additional means of analyzing existing data that would serve to either support or undermine such challenges. The rational deliberation that the hypothesis testing/weight-of-evidence approach promotes can never substitute entirely for adversarial contestation, but it can significantly narrow the gap about what is reasonably contestable. It diverts the mind away from what is most argumentative and subjective and toward what is most observable and testable, and therefore improves the agency's regulatory decisionmaking capacity.

PATH DEPENDENCY AT EPA

EPA lacks the capacity to lead such a deliberative approach, however, and the keys to understanding why can be found in its birth and early development. A recent generation of social scientists has developed the concept of "path dependence" to explain why various forms of policy regimes persist even as they become increasingly dysfunctional (Pierson 2001). An earlier generation of social scientists labeled this same phenomenon "institutionalization" (Selznick 1984). Regardless of which label we attach, the idea is that policy regimes and the agencies that administer them are products of the distinctive political and institutional environment in which they were born and subsequently nurtured in their early years. The initial decisions about regulatory principles, institutional organization, and policy design tend to persist over time because the political and bureaucratic interests that they favor mobilize politically to protect them from change. The gains that those interests realize are palpable, and therefore the energy and talent that they exert in defense of the existing regime tend to overwhelm the forces of those interests that would benefit from change (see Wilson 1973).

Thus to understand the nature of EPA's regulatory regime in general, and with regard to chemicals in particular, it is first necessary to examine the political, economic, and cultural context of the early 1970s, the era in which EPA was born and became established politically and institutionally. It is important to remember that institutionalization, or path dependency, is not a deterministic principle. The concept does not hold that transformative change is impossible, only that it is difficult. Because of the inertial pressures established at an early stage, the presumption is against change. But the presumption is rebuttable. The description of EPA's inertial bias is meant not to encourage pessimism and fatalism, but merely to show what would-be change agents are up against and suggest what sorts of political circumstances and leadership would be required to overcome the inertial presumption.

The political environment in which EPA was born had several key features that help explain its subsequent path (see Reich 1971). Most important was the prevalence in public debate of what Glendon dubbed "rights talk" (1993). The environmental movement and the politicians who identified with it were heavily influenced by the success of the Civil Rights movement of the 1960s. They recognized that posing a policy position as a right greatly enhanced its political weight and appeal. That is, if Americans had the right to a clean environment, then the standard arguments against stringent environmental policy, such as cost and unwanted governmental intrusion, paled in comparison. Although rights properly understood are never absolutes—famously, one may not cry "fire" in a crowded theater—they create a powerful presumption in favor of their exercise. For example, the Clean Air Act of 1970 prohibits an overt balancing of the public health benefits of achieving healthier air against the up-front fiscal costs incurred, thus requiring ambitious pollution reductions that would not survive a cost–benefit test (Landy 1999). Although not every environmental law proscribes such balancing, the overall commitment of EPA and its environmentalist and congressional supporters was unambiguously expressed in terms of rights, and the public absorbed that definition.

In his foreword to *Vanishing Air*, published prior to the passage of the 1970 act, Ralph Nader defined the pollution problem as the failure of the government to prosecute polluters for "harming our society's most valued rights": "Clean air is a right because it is necessary for preserving Man's essential nature: The limits that must be imposed on social and technological innovations are determined not by scientific knowledge or practical know-how but by the biological and mental nature of man which is essentially unchangeable" (Esposito 1970, *viii*).

The Naderite appeal to human nature was echoed in the congressional testimony of Sierra Club executive director Michael McClosky, who argued, "The parameters of ecological health are not negotiable. Nature has its law of limits. *Absolute results ensue when certain thresholds are crossed*, whether our political and economic institutions care to recognize them or not" (U.S. Congress 1970; emphasis in original). More notably, the chief sponsor of the proposed Clean Air Act, Senator Edmund Muskie (D-ME), adopted the rights rhetoric rather than continue to defend the existing program of state initiative (see Jones 1975). In a speech introducing his revised bill on the Senate floor, he proclaimed the right of every American to clean air:

> 100 years ago the first board of health in the United States, in Massachusetts, said this: "We believe that all citizens have an inherent right to the enjoyment

of pure and uncontaminated air." … 100 years later it is time to write that kind of policy into law … anybody in this nation ought to be able at some specific point in the future to breathe healthy air" (U.S. Congress 1974).

The National Air Quality Standards, the core of the 1970 Clean Air Act and its subsequent revisions, were explicitly based on this understanding. These standards guaranteed the *right* of any individual, regardless of existing health status, not to be harmed by breathing the air. They provided a "sufficient margin of safety" so that the health of "sensitive individuals," those with preexisting lung impairment such as asthmatics, would not be harmed. The act specifically prohibits any consideration of cost in setting the standard, as the application of the utilitarian calculus inherent in a comparison of costs and benefits would refute the rights-conferring status of the standards. A similar rights-based approach, although with very different institutional designs to achieve it, can be found in other environmental laws enacted throughout the decade, ranging from the Clean Water Act of 1972 to the Comprehensive Environmental Response, Compensation, and Liability Act (popularly known as Superfund) of 1980.

Public support for a stringent, rights-based environmental policy regime was facilitated by an equally widespread perception that the major obstacle to fulfilling environmental rights lay in the malfeasance and intransigence of corporate polluters. In this regard, the first EPA administrator, William Ruckelshaus, determined that the only viable means for establishing the credibility of the fledgling regulatory agency was to engage in a series of well-publicized enforcement actions against several corporate titans (Andrews 1999), and the early efforts of EPA were largely devoted to bringing the likes of Dow, Chevron, and Monsanto to heel (Quarles 1976; Sansom 1976). This approach had the intended effect. A spate of successful prosecutions helped establish EPA's public credibility and undoubtedly had a deterrent effect on large polluters, which feared that they would be sued if they did not quickly improve their pollution-control practices on their own. But it also served to reinforce the notion that environmental improvement could be achieved without much in the way of public sacrifice. It instilled the idea that pollution control was mainly a police action in which the "bad guys" would be reined in.

Industry responded predictably. During the rest of the decade, it closed ranks and sought to fight legal and political fire with fire. The extraordinarily rigorous hazardous waste regulations embodied in the Resource Conservation and Recovery Act of 1976 and the increased stringency of revisions of the Clean Air Act, enacted in the face of vociferous and well-financed lobbying opposition from a wide variety of business interests, attest to the failure of this approach. Industry also struggled to become as adept at using the courts to modify and ease the impact of the vast array of statutory and regulatory obligations it was saddled with as environmental-group and other public-interest lawyers had become in using litigation to toughen and strengthen those laws (Landy et al. 1990).

The Changing World of the 1990s and Now

The political, economic, and intellectual dynamics surrounding environmental matters has changed dramatically since the 1970s, yet the environmental regulatory regime has remained remarkably the same. Such is the power of path dependence. The last two decades of the twentieth century in particular saw a profound and pervasive shift in the way business and government are viewed. By the end of the 1970s, the combined impacts of the war in Vietnam, the perceived failures of Lyndon B. Johnson's Great Society, the Watergate scandal, and the energy and Iranian hostage crises had dissipated public support for government and created the conditions for widespread public receptivity to Ronald Reagan's claim that "government was the problem" (ANES 2005).

Moreover, the extraordinary growth and prosperity of information-technology companies starting in the 1980s gave a glamour and popularity to entrepreneurs and inventors that they had not enjoyed since before the Great Depression, with billionaires like Bill Gates and Steve Jobs attaining rock-star status. Unlike previous technological advances, such as nuclear power and chemical-intensive agriculture, the fruits of the information-technology revolution were almost universally considered to be sweet and nourishing. This more positive image of business in general, and high-tech business in particular, was accompanied by growing anxiety over American economic competitiveness. Globalization increasingly became a household word. Whereas earlier industry claims that environmental regulation would cause plant closings were widely seen to constitute a cry of wolf, when high-tech firms decried the impact of proposed new regulation on their ability to introduce new products, such claims were taken far more seriously (Hart 2002).

In addition, within the environmental sphere, key industries shifted their political strategy and tactics to further burnish their image. No longer did they reflexively oppose environmental initiatives. Instead, they looked for ways to bend such initiatives in a direction more favorable to them, perhaps by fashioning them such as to create barriers to entry against new competition. In deciding whether to oppose additional regulation, they also became far more sensitive to the public-relations costs of appearing environmentally unfriendly and weighed these against the costs of complying with the new regulation. To reinforce this new image, several of the most important and powerful industrial associations changed their names: the Chemical Manufacturers Association became the American Chemistry Council and the Pharmaceutical Manufacturers Association became PhRMA.

In the 1990s, environmentalism also underwent a profound change in the face of the shifting public mood regarding business and technology. The strain of "deep" environmentalism that dominated the 1970s, with its pronounced antipathy to technology and capitalism, did not disappear but increasingly shared the spotlight with advocacy for a "green" technology that would have win-win outcomes for both environmental improvement and corporate profitability. Michael Porter's "Green Gold" article was a seminal statement of this perspective, and Vice President Al Gore became its most prominent and politically influential spokesman (LaMonica 2008; Porter 1991).

The impact on the environmental movement was shown by the increasing willingness of mainstream environmental organizations to collaborate with industry, in some cases leveraging the reputations they had gained in grassroots advocacy to give corporations a seal of approval in exchange for "greener" industry practices (Bosso 2005). The first major example came in 1989, when the Environmental Defense Fund (EDF) entered into a highly publicized collaboration with McDonald's aimed at pollution prevention. In the first 10 years of the collaboration, the company replaced polystyrene containers with paper wraps and lightweight boxes made out of recycled cardboard, substituted unbleached for bleached paper carry-out bags, and made other packaging improvements throughout its supply chain (Stafford and Hartman 1996). It came as no surprise when, in 2005, EDF and the American Chemistry Council issued a Joint Statement of Principles on Nanotechnology that consisted of "several fundamental principles on which a governmental program for addressing potential risks of nanoscale materials should be premised," including a recognition of the societal benefits of nano; a commitment to discovering whether and how much risk it poses; a plea for greater government funding of nano risk research; and "an international effort to standardize testing protocols, hazard and exposure assessment approaches, and nomenclature and terminology" (ACC 2005). In September 2005, EDF and DuPont agreed to jointly establish a framework for the safe and environmentally responsible development, production, use, disposal, and recycling of nanomaterials (Bosso and Rodrigues 2006; Nano Risk Framework 2007). Although other environmental organizations were critical about EDF's approach to working with corporations, its stance reflected the views of one strand of mainstream environmentalism about how best to respond to the potential effects of emerging technologies.

The Difficulty of Changing EPA Policy

EPA has not been isolated from or immune to these broader societal and political trends. Indeed, since the mid-1990s, we have witnessed a wide variety of agency efforts to foster cooperation with industry and ease the burden of command and control regulations in laws enacted decades earlier. Among the most significant of these efforts was Project XL, which sought to use a waiver process to enable specific plants to engage in negotiations with the agency and local stakeholders to produce a plan that would give the plant greatly increased decisionmaking flexibility in return for producing environmental results that were superior to what the relevant legal and regulatory standards demanded. Another important example was the National Environmenal Performance Track (see Chapter 4), which enabled those firms with a demonstrated track record of environmental compliance to be exempted from certain onerous EPA regulatory and documentary requirements (Fiorino 2006; Marcus et al. 2002).

These efforts were laudable, even with their limitations (see Chapter 4), but they operate at the margins of the agency's activities. In critical areas of environmental regulation, the fundamentals of the old regime, rooted in statutes enacted in the 1970s, remain in place. Despite adoption of a cap-and-trade program for SO_2, the core of federal clean-air policy is still the command and control program of air-quality standards and emissions limits (Portney 2000). Water pollution is still governed by the decades-old National Pollutant Discharge Elimination System (NPDES) wastewater permitting system. Hazardous waste handling and disposal remain dictated by the famously complicated regulations emanating from the Resource Conservation and Recovery Act. Each one of these programs has come under withering attack from environmental economists who repeatedly show that the same high levels of environmental protection could be achieved at lower cost and with much higher rates of technological innovation if the inflexibility of command and control were replaced with the greater flexibility and efficiency of taxes, tradable permit systems, and other alternative regulatory schemes (Freeman 2000). Given the shifts in outlook across a wide spectrum of environmental matters at all levels of government, this statutory-rooted inertia may seem surprising, but it is a testament to the inherent difficulties the American political system puts in the way of systematic policy change (Baumgartner and Jones 1993; Bosso 1987).

Industry would welcome a regulatory order that was more attuned to risk–benefit and cost–benefit considerations (see Graham and Wiener 1995), but for all the improvements that have occurred in its image, its word is still not trusted by the American public (Pew Research Center 2003). The periodic surfacing of public health threats such as spinach contaminated with E. coli, salmonella-tainted peanut butter, and lead paint found on toys imported from China remind the public of the risks of putting too much trust in the capacity of producers and retailers to police themselves and their suppliers (CDC 2006; Moss 2009; Story 2007).

For its part, the environmental movement remains divided about how far to go to accommodate industry. Melnick (1992) argues that in the 1970s in particular, the movement gained much of its strength and staying power by pursuing a political strategy of "disappointed expectations." In this view, environmentalists pushed EPA to adopt standards that they knew the agency could not meet, then abetted the agency's granting of time extensions to industries when the standards were in fact not met, and subsequently criticized the agency for caving in to industry. Environmental groups thus preserved their commitment to environmental rights in principle while making practical concessions. But the concessions themselves had political value, because they were made to appear as evidence of the dilatory nature of the agency and the foot-dragging of industry. Melnick provides the concrete example of the Reagan administration's effort to enforce Clean Air Act noncompliance deadlines by cutting off highway funds to noncomplying cities. Instead of backing this effort, the environmental lobby sided with the urban governments in fighting for a stretch-out of the deadlines, even as they criticized the administration for its overall enforcement of the act (Melnick 1992).

But times have changed, and the diversity of philosophy and tactics within organized environmentalism are more apparent (Bosso 2005). Such differences among environmentalists regarding nanotechnology are exemplified by the close working relationship that has developed between EDF and DuPont, on the one hand, and calls by Greenpeace and Friends of the Earth for an immediate moratorium on the release of nanomaterials and products, on the other (Johnston et al. 2007; FOE 2006). Industry, as

Cary Coglianese notes in Chapter 4, is fragmented into disparate sectors and exhibits no clear consensus on future directions. Caught in the middle, EPA has little leverage of its own to decide the matter. It could propose changes in TSCA, for example, but Congress is unlikely to pass them, even with a new president, for many of the same political reasons that have defined environmental regulatory policymaking going back to the origins of the agency nearly 40 years ago.

NANO AS A GRAND BARGAIN

The great danger in stressing the obstacles to policy change is that it makes such change appear impossible, when in fact changes on a grand scale do indeed occur, albeit sporadically (Baumgartner and Jones 1993). They occur when the inertial dynamics that normally dominate those policy spheres are transcended by a higher politics. The use of the term "higher" is not meant as an ethical commentary, but as an indication that the issue at stake has been temporarily displaced from its normal political milieu of congressional subcommittees, agency officials, and interest groups and elevated to a different political plane that introduces different political actors with different political motives. The ordinary politics of the issue are not altered; rather, they are displaced (Schattschneider 1960).

A clear example is the 1986 Tax Reform Act. The lobbyists seeking tax breaks for their clients and the members of Congress seeking campaign contributions from the lobbyists did not disappear; rather, the tax policy debate was moved to a higher plane involving direct public appeals for reform by an influential senator and a public commitment to reform by the president. Having made such pledges, both Senator Bill Bradley (D-NJ) and President Ronald Reagan, a Republican, fought to keep tax reform at the top of the congressional agenda. The high level of visibility and urgency with which they and their allies were able to endow the issue vastly diminished the political power of the interest groups and members of Congress who dominated tax policy under normal, low-visibility conditions. Once this higher plane was reached, the bargain was relatively simple: liberals obtained a reduction in tax subsidies for corporations, conservatives obtained lower and fewer rates (Conlan et al. 1995).

Environmental regulation of nanotechnology, despite the myriad tech-nical difficulties it poses, might soon provide a good opportunity for this sort of grand bargain. The very attributes that make it politically problematic for industry may encourage both sides to take political risks they would normally avoid. Nanotechnology puts both environmentalists and industry off their accustomed game, because neither side can ignore the great benefits it promises or the great need for government validation to avoid public skepticism or worse. Leaders in the environmental movement see the inherent dangers both of being too pro-nano, because they do not yet have a grasp of the magnitude of the risks nano poses, or being reflexively anti-nano, and thus being perceived as perversely seeking to deprive society of the enormous benefits nanotechnology promises, including great environmental ones. For its part, industry has to weigh the benefits of escaping overly stringent regulation against its powerful need to obtain government validation that nano is safe.

The precedent that haunts nano proponents is genetically modified (GM) food, a highly promising technology whose growth was stunted by the high level of public fear it engendered, particularly in Europe (Sandler and Kay, 2006). To proponents of technological innovation in particular, the fate of GM food is a sad reminder of how quickly the public can turn against an emerging technology in the face of charges, well founded or not, about the damage it may do to the environment. Thus far, environmentalists remain ambivalent about whether to treat nano as an environmental threat in the vein of GM food or embrace it as a superior means for solving a wide array of environmental and agricultural problems. Industry is torn between seeking to avoid cumbersome and expensive regulatory burdens and beseeching government to validate nano's environmental soundness as a shield against GM-like environmental attacks.

These ambivalences could set the stage for a grand political bargain: environmentalists support government validation of nanotechnologies in exchange for industry willingness to take on a far greater amount of preproduction environmental testing and data-sharing. Both sides thus obtain what they need the most. Industry obtains a government seal of approval for specific products and processes and may improve its overall public image in the process. Environmentalists gain the possibility of improved forms of environmental mitigation and escape the political trap of being branded as anti-technological Luddites. Both sides also make sacrifices. Industry must endure public probing into matters it would greatly prefer to keep proprietary. Environmentalists give up the ability to reflexively stir up fear about the risks of an emerging technology as a tool for mobilizing political support and financial contributions.

As the tax reform example demonstrated, bargains of a grand scale and scope do not just happen; they require political leadership, preferably at the presidential level. Nor do they need to await a crisis to push the protagonists out of their comfort zone. There was nothing remarkable about the economic circumstances of the United States in 1986, nor was the public demanding tax reform. To overcome the ordinary inertial forces that pervade the Washington policymaking process, creativity can substitute for crisis. A new president convinced of the importance to the nation's international competitiveness and future prosperity of removing hindrances to the manufacturing and sale of nano products could challenge industry, environmentalists, and congressional leaders to accept such a grand bargain. Environmentalists would have to accept a risk-based approach to assessing the dangers that nano products and processes pose to the environment more or less along the lines outlined in this chapter. In return, industry would have to live up to its promise to provide the data EPA needs to make such designations.

This chapter has examined the most serious forms of incapacity regarding EPA's regulation of nanotechnology, as well as an approach to risk assessment that appears to constitute the most promising form of improvement. Once we understand the political sources of this incapacity, it becomes clear that the solution lies in the form of politics capable of pushing meaningful reform to a higher political plane and offering the possibility of a grand bargain. Full recognition of what accounts for EPA's regulatory incapacity thus points the way to overcoming it.

REFERENCES

ABA (American Bar Association). 2006. *Regulation of Nanoscale Materials under the Toxic Substances Control Act.* Chicago, IL: ABA, Section of Environment, Energy, and Resources.

ACC (American Chemistry Council). 2005. *Environmental Defense and the Nanotechnology Panel of the American Chemistry Council: Joint Statement of Principles.* Arlington, VA: ACC.

Andrews, Richard N. L. 1999. *Managing the Environment Managing Ourselves: A History of American Environmental Policy.* New Haven, CT: Yale University Press.

ANES (American National Election Studies). 2005. Trust in Government Index, 1958–2004. ANES Guide to Public Opinion and Electoral Behavior. www.electionstudies.org/nesguide/toptable/tab5a_5.htm (accessed March 8, 2009).

Baumgartner, Frank R., and Jones, Bryan D. 1993. *Agendas and Instability in American Politics.* Chicago, IL: University of Chicago Press.

Bergeson, Lynn L. 2007. The EPA's Toxic Substances Control Act: What You Must Know. www.smalltimes.com (accessed February 28, 2009).

Bosso, Christopher. 1987. *Pesticides and Politics: The Life Cycle of a Public Issue.* Pittsburgh, PA: University of Pittsburgh Press.

———. 2005. *Environment Inc.: From Grassroots to Beltway.* Lawrence, KS: University Press of Kansas.

Bosso, Christopher, and Ruben Rodrigues. 2006. Emerging Issues, New Organizations: Interest Groups and the Making of NanoTechnology Policy. In *Interest Group Politics*, 7th ed., edited by Allen J. Cigler and Burdett A. Loomis. Washington, DC: CQ Press, 366–88.

CDC (Centers for Disease Control and Prevention). 2006. Multi-State Outbreak of *E. coli* O157:H7 Infections from Spinach. www.cdc.gov/ecoli/2006/september/ (accessed March 4, 2009).

Conlan, Timothy J., David R. Beam, and Margaret T. Wrightson. 1995. Policy Models and Political Change: Insights from the Passage of Tax Reform. In *The New Politics of Public Policy*, edited by Marc K. Landy and Martin A. Levin. Baltimore, MD: John Hopkins University Press, 121–42.

Corrosion Proof Fittings v. EPA. 1991. U.S. Court of Appeals for the Fifth Circuit. 947 F.2d 1201.

Davies, J. Clarence. 1996. *Comparing Environmental Risks: Tools for Setting Government Priorities.* Washington, DC: Resources for the Future.

———. 2007. *EPA and Nanotechnology: Oversight for the 21st Century.* Washington, DC: Woodrow Wilson International Center for Scholars.

DeLisi, V. M. Jim. 2009. Revisiting the Toxic Substances Control Act of 1976. Testimony given to the Subcommittee on Commerce, Trade, and Consumer Protection, of the Committee on Energy and Commerce, U.S. House of Representatives, 111th Congress. On behalf of the Synthetic Organic Chemical Manufacturers Association. February 26.

Denison, Richard A. 2009. Ten Essential Elements of TSCA Reform. *Environmental Law Reporter* 39: 10022–28.

EDF (Environmental Defense Fund). 2006. *Why "Existing Chemical SNURs" Won't Suffice to Protect Human Health and the Environment.* Washington, DC: Environmental Defense Fund.

EPA (Environmental Protection Agency). 2005. Guidelines for Carcinogen Risk Assessment. *Federal Register* 51 (185): 33992–34003.

———. 2007. *Toxic Inventory Status of Nanoscale Substances—General Approach.* Washington, DC: Environmental Protection Agency.

Esposito, John C. 1970. *Vanishing Air.* New York, NY: Pantheon.

Fiorino, Daniel. 2006. *The New Environmental Regulation.* Cambridge, MA: MIT Press.

Fletcher, Susan R., Claudia Copeland, Linda Luther, James E. McCarthy, Mark Reisch, Linda-Jo Schierow, and Mary Tiemann. 2008. *Environmental Laws: Summaries of Major Statutes Administered by the Environmental Protection Agency (EPA).* Washington, DC: Congressional Research Service.

FOE (Friends of the Earth). 2006. Nanomaterials, Sunscreens and Cosmetics: Small Ingredients, Big Risks. www.foe.org/pdf/nanocosmeticsreport.pdf (accessed June 1, 2009).

Freeman, A. Myrick III. 2000. Water Pollution Policy. In *Public Policies for Environmental Protection*, edited by Paul Portney. Washington, DC: Resources for the Future.

GAO (U.S. General Accounting Office). 1994. *Toxic Substances Control Act: Legislative Changes Could Make the Act More Effective.* GAO/RCED-94-103: 18. Washington, DC: U.S. General Accounting Office.

GAO (U. S. Government Accountability Office). 2005. *Options Exist to Improve EPA's Ability to Assess Health Risks and Manage Its Chemical Review Program.* GAO-05-458. Washington, DC: U.S. Government Accountability Office.

Glendon, Mary Ann. 1993. *Rights Talk: The Impoverishment of Political Discourse.* New York, NY: Free Press.

Graham, John D., and Jonathan Baert Wiener. 1995. *Risk versus Risk: Tradeoffs in Protecting Health and the Environment.* Cambridge, MA: Harvard University Press.

Hart, David. 2002. High Tech Learns to Play the Washington Game. In *Interest Group Politics*, 6th ed., edited by Allan J. Cigler and Burdett A. Loomis. Washington, DC: CQ Press, 293–313.

Johnston, Paul, David Santillo, John Hepburn, and Doug Parr. 2007. *Nanotechnology Policy & Position Paper.* Washington, DC: Greenpeace.

Jones, Charles O. 1975. *Clean Air: The Politics and Policies of Pollution Control.* Pittsburgh, PA: University of Pittsburgh Press.

LaMonica, Martin. 2008. Al Gore: Business Is Ahead of Government on Climate Change. news.cnet.com/8301-11128_3-9898088-54.html (accessed March 4, 2009).

Landy, Marc K. 1999. Environmental Policy and Local Government. In *Dilemmas of Scale in American Democracy*, edited by Martha Derthick. Cambridge, UK: Cambridge University Press, 227–60.

Landy, Marc K., Marc J. Roberts, Stephen R. Thomas, and U.S. EPA. 1990. *Asking the Wrong Questions.* New York, NY: Oxford University Press.

Marcus, Alfred A., Donald A. Geffen, and Ken Sexton. 2002. *Reinventing Environmental Regulation: Lessons from Project XL.* Washington, DC: Resources for the Future.

Melnick, R. Shep. 1992. Pollution Deadlines and the Coalition for Failure. In *Environmental Politics: Public Costs, Private Rewards*, edited by Michael S. Greve and Fred L. Smith, Jr. New York, NY: Praeger, 98.

Moss, Michael. 2009. Safety Net Missed Problems at Peanut Plant. *New York Times*, February 9, A1.

Nano Risk Framework. 2007. Executive Summary. www.nanoriskframework.com/content.cfm?contentID=6498 (accessed March 8, 2009).

Pew Research Center. 2003. The 2004 Political Landscape. Part 7: Business, Government, Regulation and Labor. Pew Center for the People and the Press: Survey Reports. people-press.org/report/?pageid=756 (accessed March 4, 2009).

Phillips, Melissa Lee. 2006. Obstructing Authority: Does EPA Have the Power to Ensure Commercial Chemicals Are Safe? *Environmental Health Perspectives* 114 (12): A708.

Pierson, Paul. 2001. *The New Politics of the Welfare State*. New York, NY: Oxford University Press.

Porter, Michael. 1991. America's Green Strategy. *Scientific American* 264 (April): 168.

Portney, Paul R. 2000. Air Pollution Policy. In *Public Policies for Environmental Protection*, edited by Paul Portney. Washington, DC: Resources for the Future, 112.

Quarles, John. 1976. *Cleaning Up America: An Insider's View of the Environmental Protection Agency*. Boston, MA: Houghton Mifflin, 117–18.

Reich, Charles. 1971. *The Greening of America*. New York, NY: Bantam.

Rhomberg, Lorenz. 2007. *Hypothesis-Based Weight of Evidence: An Approach to Hazard Identification and to Uncertainty Analysis for Quantitative Risk Assessment*. Cambridge, MA: Gradient Corporation, and Arlington, VA: Aerospace Industries Association.

Sansom, Robert L. 1976. *The New American Dream Machine: Toward a Simpler Lifestyle in an Environmental Age*. Garden City, NY: Anchor Press, 24–25, 43.

Sandler, Ronald, and W. D. Kay. 2006. The GMO-Nanotech (Dis)Analogy? *Bulletin of Science, Technology & Society* 26 (1): 57–62.

Schattschneider, E. E. 1960. *The Semisovereign People: A Realist's View of Democracy in America*. New York, NY: Harcourt Brace College Publishers.

Selznick, Philip. 1984. *Leadership in Administration*. Berkeley, CA: University of California Press.

Stafford, Edwin R., and Cathy L. Hartman. 1996. Green Alliances: Strategic Relations between Businesses and Environmental Groups. *Business Horizons* 39 (March–April): 50–59.

Story, Louise. 2007. Lead Paint Prompts Mattel to Recall 967,000 Toys. *New York Times*, August 2.

U.S. Congress. House. Subcommittee on Conservation and Natural Resources, Committee on Government Operations. 1970. *The Environmental Decade: Action Proposals for the 1970's*. 91st Cong., 2nd sess., February 2–6; March 13; April 3.

U.S. Congress. House. 1976. *Toxic Substances Control Act (TSCA)*. 94th Cong., 2nd sess. Public Law (PL) 94–469. www.osha.gov/dep/oia/whistleblower/acts/tsca.html (accessed March 4, 2009).

U.S. Congress. Senate. Committee on Public Works. 1974. *A Legislative History of the Clean Air Act of 1970*. Debate on S. 4358, September 21, 1970. 93rd Cong., 2nd sess., 1974, 220, 227.

Wilson, James Q. 1973. *Political Organizations*. New York, NY: Basic Books.

Wright, Michael J. 2009. Revisiting the Toxic Substances Control Act of 1976. Testimony before the House Subcommittee on Commerce, Trade, and Consumer Protection. On Behalf of the United Steelworkers. February 25.

CHAPTER 6

NANOTECHNOLOGY AND THE EVOLVING ROLE OF STATE GOVERNANCE

Barry G. Rabe

Most analyses of alternative approaches to American governance of the potential environmental risks posed by nanotechnology have presumed a dominant, and perhaps exclusive, role for the federal government. This reflects the national scope of anticipated development and use of nanotechnology, as well as the possibility of some future form of multinational or international collaboration in which the federal government would be a lead player. As a result, much existing analysis about possible responses to nanotechnology's environmental effects focuses on whether to rely on established federal legislation, such as the Toxic Substances Control Act (TSCA), rather than new legislation expressly tailored to unique challenges raised by nanotechnology. In turn, governance questions tend to focus on the viability of one federal government unit, likely the Environmental Protection Agency (EPA), taking the lead role in any policy implementation, as opposed to some nano-specific interagency consortium or network that transcends jurisdictional boundaries among the many federal agencies and departments with an expressed interest in nanoscale research, commercialization, and oversight.

Yet the evolving American experience in nanotechnology suggests an even more complex policy process, one embedded in the realities of U.S. federalism. As in many other areas of environmental protection predating the emergence of concerns about nanoscale particles and applications, the federal government is not likely to hold exclusive power. It will, instead, share governance responsibilities with states and even localities. This pattern of shared responsibility has held for generations and, in fact, may have intensified during the 2000s. Indeed, in the past decade, states have taken surprisingly active roles in areas such as climate change and stem cell research and development that conventionally would have been deemed the province of Washington, D.C. (Mintrom 2009) It remains unclear how governance duties in these rapidly emerging areas will be allocated over coming years, particularly given the significant economic downturn in 2008–2009. But it is likely that some form of shared or multilevel governance will endure, even if the federal government enacts new legislation or pursues multinational approaches.

At one level, nanotechnology presents a classic challenge for states. They must square their abiding interest in promoting internal economic development against any environmental protection concerns. Traditionally, states were thought to focus almost exclusively on the former and do, in fact, pursue a wide range of strategies to foster development, whether designed to retain current investment or create incentives for new initiatives. But they have increasingly turned toward environmental protection as a central focus as well, reflecting growing public concern about environmental quality, and often in response to issues that are particularly salient for a given jurisdiction. Nanotechnology presents a somewhat similar tension for states, as they explore how best to maximize the likelihood that they will become centers for any related economic benefits accruing from its expansion, while also guarding against any environmental risks. At the same time, nanotechnology presents an added wrinkle in that it may pose environmental threats but its very use may also promote environmental improvement, such as superior technologies for remediating contamination or increasing energy efficiency. This blending of environmental "bad" with "good" has begun to emerge in at least a few state cases, although the economic development imperative remains dominant in most contexts.

This chapter presents an early attempt to consider nanotechnology as a challenge for all levels of government within a federal system such as the United States, but with a particular focus on the evolving role of states. It begins with an overview of recent developments in American intergovernmental relations, including important factors that gave states such a significant role in many areas of environmental policy. The analysis also provides reference to other environmental issues, most notably climate change, that once appeared headed toward dominance by national and international governments but now find states as central policy players. This leads to a discussion of possible incentives for states to become involved in nanotechnology, reflecting perceptions of economic development benefits that may result from state government efforts to promote nanotechnology development and concerns about potential environmental health risks to citizens. This portion of the analysis includes a review of existing state policy. It considers a range of economic development strategies but devotes particular attention to the cases of California, Massachusetts, Minnesota, and Wisconsin, which have increasingly added an environmental protection focus. The chapter concludes with a discussion of the challenges and opportunities present when placing nanotechnology into an intergovernmental context.

THE ENDURING AND EXPANDING STATE PRESENCE

State governments face considerable constraints, many established through legal and political interpretations of the U.S. Constitution that expanded federal government authority during the latter two-thirds of the twentieth century. Such limitations on state authority are particularly severe in instances that would clearly involve them in international relations or entail any effort to promote internal economic development by precluding the import of goods from other states. A number of areas of public policy have essentially operated on a unitary basis, ranging from most aspects of national defense to the provision of senior citizen pensions through the Social Security program. Yet even these highly centralized areas continue to retain some role for state government. In military affairs, the expanding role of state-based military reserves, such as National Guard units, has been increasingly evident in recent years in U.S. interventions in Afghanistan and Iraq. In Social Security, the medical care insurance component of the federal safety net was expanded in 2004 to include coverage of many prescription drugs for senior citizens. Although generally portrayed as a federal policy, this expansion of the Medicare Part D program involved exhaustive federal and state negotiations, reflecting long-standing commitments by many states to provide such support, as well as major differences in state and regional allocation of prescription drugs (Derthick 2007). Even the many moves to create a centralized set of policies and institutions to address homeland security in the aftermath of the September 2001 attacks retain vast roles for state and local authorities (Kettl 2007).

States also face legal and political constraints in many areas of environmental protection, many imposed during the expansion of federal environmental regulatory activity between 1969 and 1974 (Jones 1975). Even the most centralized pieces of federal environmental policy, however, generally do not exclude a state role. In the case of the Superfund program, established by the Comprehensive Environmental Response, Compensation and Liability Act of 1980 to address the most egregious cases of abandoned hazardous waste dumps, federal legislation vested enormous regulatory authority with EPA. But states actively jockey to secure their share of Superfund support and work closely with regional offices of the federal agency on the details of major cleanup decisions. More than half of the states also established parallel programs of their own to fill gaps when federal resources are invariably limited (Nakamura and Church 2003). In the case of the TSCA, perhaps the most centralized piece of federal environmental legislation, and widely noted as a possible player in regulating nanotechnology, states continue to carve out a range of policies to address the registration and regulation of toxic chemicals within their jurisdictions, including a wide range of state programs designed to reduce use and prevent exposure.

Most other areas of environmental protection involve formal sharing of authority, even in instances where a federal statute exists. This dynamic often entails some semblance of formal federal delegation of responsibility, such as permitting and related enforcement actions that give states enormous latitude to set their own priorities. At the same time, core areas such as air- and water-pollution control leave considerable room for states to innovate, whether setting standards that are more rigorous than those of the federal government or experimenting with new methods that are intended to promote environmental improvement alongside administrative efficiency (Fiorino 2006; Scheberle 2004). Not surprisingly, states tend to focus heavily on those issues of particular concern to them, whether air quality in California or Great Lakes water protection and diversion in Michigan.

If anything, the extent of American state engagement in many areas of environmental protection has expanded and intensified during the past 20 years as a result of a combination of factors at both the federal and state levels. At the federal level, the pace of either reauthorizing existing legislation or enacting new legislation in response to evolving challenges has slowed markedly since the explosion of federal environmental legislative activity in the 1970s. Congress has struggled to reach consensus on a wide range of environmental issues, despite varied patterns of partisan control of the House and Senate (Kraft 2010). Modest reforms that largely entail necessary updates and refinement have consistently proven contentious, much less larger challenges such as bolstering the very limited TSCA (Davies 2006). In turn, initial federal responses to new areas such as climate change or nanotechnology have been elusive.

Congressional stasis has created policy voids that, in many instances, state governments have attempted to fill. Moreover, the federal government has lagged badly in investing in its regulatory infrastructure, even as new issues and challenges have been added to the agendas of agencies such as EPA. As environmental policy scholar J. Clarence Davies has noted, EPA has fewer staff now than it did a decade ago, and it has a smaller overall budget in constant dollars than in the early 1970s (Davies 2007; Davies and Rejeski 2007). Such conditions leave scant room for intra-agency innovation or the kind of agency-based policy entrepreneurship that has been so evident at the state level.

In response to this apparent absence of federal leadership, states have proven increasingly willing to invest their own resources into expanding agency capacity and launching new policies via legislation or gubernatorial executive orders. This pattern varies markedly by individual state, but the overall trend is one of considerable state preparedness to challenge federal interpretation of existing legislation or simply to fill gaps where they appear. State policy to address greenhouse gas reductions may be the most visible manifestation of this phenomenon, but it is also evident in such areas as reducing air emissions of mercury, establishing land-use controls to reduce urban sprawl, and recycling and reuse of waste materials from computer products (Rabe 2007). State government authority has clearly expanded in many areas relevant to environmental protection in recent decades and has generally been successful in moving beyond earlier governance styles that allowed for undue influence by regulated parties in policy decisions (Teske 2004). This has reflected a growing sense that states tend not to see aggressive environmental protection as contrary to their economic self-interest. Indeed, many state policy initiatives directly link the need to pursue environmental protection with fostering economic development, whether through maintaining a high quality of life and an attractive climate for investment or developing technologies and expertise commensurate with national or even international leadership in a more environmentally oriented society. It remains unclear how mounting state fiscal problems in

2009 will influence these efforts, though some states appear to be sustaining their commitments in the hope of diversifying their economies.

At the same time, the expansion of state agency capacity in areas such as environmental protection, energy, public health, and natural resources has created new venues for policy innovation and entrepreneurship (Rabe 2004). Many states have expanded their staffs and modified lead agencies and departments as new environmental challenges arise, creating opportunities for innovation that may not have been available to federal-level counterparts. Unlike the federal government, many states also feature the option of direct democracy through referenda and initiatives, offering citizens an increasingly popular way to advance new environmental protection legislation or raise funds through bond issuance (Guber 2003). Finally, states tend to have a relatively large number of top elected officials, such as an attorney general and a treasurer, who are not beholden to the governor and may even belong to a different political party. Such key state offices increasingly serve as additional and independent venues for environmental policy innovation.

THE CASE OF CLIMATE CHANGE

The evolving American experience with climate-change policy illustrates the penchant for substantial state-level engagement, even in an area in which most conventional analysis anticipated an exclusively federal and international governance regime (see Rabe 2004, 2008). The federal government has been active in international diplomacy on this issue for more than two decades, and Congress held more than 400 hearings on various aspects of climate change between 1975 and 2008. But all of this activity in this period translated into very modest federal policy steps, primarily funding for research in the natural and physical sciences related to climate change and a sequence of voluntary reduction programs. Many legislative proposals that could reduce greenhouse gases, including carbon tax and cap-and-trade programs, have been regularly introduced in recent sessions of Congress. Few, however, have received serious consideration in one chamber, much less reached the president's desk for possible signing. The arrival of the 111th Congress and the presidency of Barack Obama signaled the possibility of substantial expansion of the federal role and yet virtually each policy option under consideration would have to address entrance into a policy area that has since been dominated by state policies that are moving into full implementation.

States were hardly obligated to fill this void, but a growing number of them chose to do so for a range of particular reasons. As of late 2009, 29 states representing each region of the nation had enacted renewable portfolio standards, which mandate a continual increase in state electricity supply that must come from noncarbon sources. For example, Arizona is now committed to attain 20 percent of its electricity from renewable sources by 2020, and New York has pledged to derive 25 percent by 2013. Twenty-three states were formally involved in developing regional carbon-emissions trading programs, and 14 had officially embraced California's 2002 legislation to establish significant carbon-emissions reductions for vehicles. A growing number of states established statewide emissions reduction commitments, with some reaching well beyond the end of the current decade and setting some of the most ambitious targets of any government in the world. Other state policies address differing dimensions of the greenhouse-gas challenge, such as capturing methane from solid-waste landfills, reforming agricultural and forestry practices, or heightening carbon sequestration. Collectively, nearly half of all states now have two or more significant climate policies in place, and these appear to have contributed at least somewhat to the surprisingly slow rate of American greenhouse-gas emissions growth over the past five years (Rabe 2008).

States have pursued these policies for multiple reasons, although the convergence of factors varies from case to case. Many states began to experience early indicators of the likely impact of climate change and decided to take unilateral action both to reduce their contribution to the problem and possibly stimulate other jurisdictions to act. States with significant amounts of coastal land—Florida, New Jersey, and Virginia in particular—are increasingly sensitive to the threats of intensified storms and rising sea levels and based a series of 2008 and 2009 policy initiatives in large part on this threat. Other regions face different challenges that state leaders and citizens attribute to climate change, ranging from drought and water scarcity to elevated temperatures posing heat-island threats in large cities.

Additionally, many states view climate change as posing an economic development opportunity through cultivation of skills, technologies, or resources that may be indigenous rather than requiring import (Rabe 2004). A major push behind the rapid expansion of interest in promoting renewable energy has been the idea of building internal capacity to tap into "homegrown" energy sources. An increasing number of states have begun to actively compete in a race to the top to demonstrate national and global leadership on climate change, and thereby position themselves advantageously for future responses to this issue. This feature has proven particularly attractive to elected officials such as Governor Arnold Schwarzenegger (R-CA), who has repeatedly championed his state's ambitious greenhouse-gas reduction efforts as promoting environmental protection as California develops new technologies and skills for a carbon-constrained economy.

The overall pattern of state involvement is highly uneven, however, ranging from those like California, which are prepared to pursue virtually every possible policy imaginable, to states such as Mississippi and Tennessee, with essentially no significant engagement on the issue to date (Rabe 2008).

There is no indication as of yet that nanotechnology will follow climate change with this level or variability of state-government engagement. It may indeed provide considerable economic development opportunities for states linked with taking lead roles on environmental protection issues related to nanotechnology development. But in the case of climate change, many states were building on existing policies and areas of expertise, such as promoting energy alternatives or extending established emissions-trading experience to carbon dioxide. It is not at all clear that many states have such an extensive body of policies and skills that would readily translate into governing nanotechnology, much less that this new technology will ever attain the saliency of climate change as an environmental policy concern (Davies 2007). Nonetheless, the climate case illustrates the possibility of reversing the conventional wisdom that presumes federal-level domination of a policy arena that relegates states to the sidelines. It further raises the question of whether a similar bottom-up pattern might emerge in nanotechnology.

STATES AND STRATEGIC OPTIONS FOR NANOTECHNOLOGY POLICY

State government decisions on whether to become involved in some aspect of nanotechnology policy may be heavily influenced by such factors as their potential for developing significant nanotechnology manufacturing sectors and the internal capacity within state agencies to develop and implement guiding policies. At least four broad options emerge that may drive strategic choice. For those states that have, or promise to develop, large nanotechnology sectors, interest in both promoting this activity and safeguarding against any environmental risks or perceptions of such risks may be heightened. But for states with little or no such development and no near-term prospects for nanotechnology sector expansion, it is far less likely that they will invest resources in either nanotechnology promotion or regulation. States also vary markedly in their capacity to address cutting-edge technologies and their potential environmental consequences. Policy development opportunity for prospective entrepreneurs may be substantial in states with considerable agency staff expertise related to chemical regulation and other areas that might transfer directly into the burgeoning field of nanotechnology. But such opportunity may be greatly diminished in those states that have more modest agency capacity or legislative constraints on agency initiative. Active policy engagement is perhaps most likely in those cases where both nanotechnology development and policy development capacity are high, most evident thus far in California, Massachusetts, Minnesota, and Wisconsin, discussed in more detail in a later section of this chapter. Far more states appear to fall into the category of high nanotechnology development but only modest policy capacity, at least to date. Still others give little evidence of any engagement on this issue from either an economic development or environmental protection perspective.

The Economic Development Emphasis

States have long searched for ways to promote economic development within their boundaries by supporting established firms or using a range of policy tools to foster new ones. Economic development options for states include tax incentives, start-up and incentive funding, research support, assistance in securing environmental and related permits, and state government guarantees to purchase newly manufactured products. States may work cooperatively with local governments that serve as the hosts of laboratories or manufacturing facilities. Just as states are exploring a range of policies designed to promote climate-friendly technology and skills that might redound to their long-term economic advantage, it is increasingly evident that they are developing numerous methods to sustain and promote nanotechnology at home (NNI Workshop 2003). At one level, this is already a relevant concern in that every state hosts at least one organization that is engaged in nanotechnology. A 2007 survey by the Woodrow Wilson International Center for Scholars concluded that 10 states had particularly large nanotechnology sectors in operation, based on the number of participating companies, universities, and laboratories: California, Massachusetts, New York, Texas, Pennsylvania, Michigan, New Jersey, Illinois, Florida, and Ohio (Keiner 2008).

At least 20 state governments have made some tangible commitment to "invest in nanotechnology infrastructure," although a number of these programs are modest in scope and little is known about their impact

(IRGC 2005, *104*; Sa et al. 2008). As Davies has noted, the state programs are "essentially part of each state's economic development strategy." He finds little evidence of state policy activities that transcend a development focus and concludes that states may face disincentives in taking regulatory action, noting, "New companies might not be encouraged to settle in a state that is looking at the harm that nano might do" (Davies 2007, *41*). I will revisit this question of whether states may be beginning to move beyond a purely developmental role after first more fully delineating the kinds of promotional policies states are putting into operation.

A 2006 survey of nanotechnology policies in the various states by the National Conference of State Legislatures (NCSL) confirms that a sizable number have made a significant economic development commitment to nanotechnology through legislation, including several smaller states with limited fiscal resources (NCSL 2006). Arkansas, for example, has enacted legislation that establishes a tax credit for money invested in developing nanotechnology; Kansas has enacted several statutes that formally make nanotechnology eligible for a number of possible state funding sources, such as the Bioscience Research and Development Voucher Program. The NCSL survey, as well as evidence gathered from other sources, suggests that larger states tend to have more expansive programs, committed to the creation of not only new funding opportunities, but also supportive infrastructure for technology development. Brief summaries follow on four of these state cases, those of New York, Texas, Pennsylvania, and Virginia. A subsequent section focuses on California, Massachusetts, Minnesota, and Wisconsin, which appear to be the most active in terms of simultaneously pursuing economic development opportunities and beginning to engage more of the environmental protection side of nanotechnology policy.

New York

The state of New York has taken a number of steps to attempt to provide support funding for new nanotechnology projects, including legislation that authorized the New York State Office of Science, Technology and Academic Research (NYSTAR) to give financial support to nanotechnology firms. NYSTAR provided nearly $100 million in nanotechnology research and development support in fiscal years 2005 and 2006. But it appears to have taken a regional focus as well, with heavy concentration on the Albany area and the State University of New York campus located in the capital. This includes extensive state support for the development of Albany NanoTech, which serves as a consortium to link private investors with researchers based at the university and in state government. Albany NanoTech is located in a $1 billion state-funded complex, which includes the New York State Center of Excellence in Nanotechnology and Nanoscience, the New York State Center for Advanced Technology in Nanomaterials and Nanoelectronics, the regional research unit of the Semiconductor Industry Association, and the College of Nanoscale Science and Engineering (NNI Workshop 2003). In 2005 alone, the Albany region received $1.9 billion in pledges from a private firm to build a new semiconductor fabrication plant and conduct nanotechnology research, supplemented by $150 million in state funds (BRTFN 2005). This region of the state has a long-standing history of involvement in related technologies, most notably semiconductors.

Texas

Texas has long aspired to national leadership in nanotechnology development, reflected in a large number of firms pursuing research and development. Many of these operations were spin-offs from the state's extensive investment in research capacity at universities around Texas. The state largely relied, however, on its "siren song of lower operating costs and access to unique, Texas-based infrastructure resources" (NNI Workshop 2003, 57). Legislation enacted in 2005 required the state's main economic development unit, the Texas Economic Development and Tourism Office, to "coordinate state efforts to attract, develop, or retain technology industries in this state in certain sectors, including nanotechnology" (NCSL 2006, 2). Additional legislation authorizes financing for a product of small business to "give preference to nanotechnology" and defines nanotechnology as an "emerging technology industry," (NCSL 2006, 2) thereby making it eligible for other sources of state financial support. This definitional change is significant in that it has allowed funding from the Texas Emerging Technology Fund for a range of nanotechnology projects. The state also has created programs that can offer loans to small or new firms pursuing nanotechnology and established the Nanoelectronics Workforce Development Initiative to expand training opportunities (Smith and Bosso 2007). In many respects, this drive parallels Texas efforts to use state policy to actively promote renewable energy, which has targeted the state's vast supply of reliable wind power and made it the largest generator of energy from this source in the nation.

Pennsylvania

Pennsylvania has a long-standing history of attempting to promote technology-based economic development, reflecting in part a continuing concern about facilitating a transition from its declining base of heavy manufacturing. Much of this work began in the 1980s, with a focus on developing distinct research hubs in the vicinity of major universities in the commonwealth. The shift into nanotechnology began in the late 1990s, with a series of modest grants to support start-up projects that expanded in subsequent years, many of which were directly linked to existing technology development programs. This activity reached a new level in 2005, with the creation of the Pennsylvania Initiative for Nanotechnology (PIN). The establishment of PIN followed a 2003 study commissioned by the secretary of community and economic development, which proposed an integration and expansion of existing state nanotechnology efforts and deemed Pennsylvania poised to become a major player in this arena. Governor Edward Rendell embraced the report and viewed PIN as "positioning the Commonwealth to be a national force in this building wave of development." PIN is intended to provide a direct focus on nanotechnology, responsible for $11.1 million in funding in 2006. According to the first scholarly analysis of PIN, "The state sustained ongoing programs that evolved from the bottom up, on a decentralized (and sometimes internally competitive) fashion," but also made state funding in this area, "at least nominally, part of a statewide initiative" (Sa et al. 2008, 10).

Virginia

Virginia has attempted to support nanotechnology development through the creation of supportive institutions, including the first cabinet-level secretary post that expressly has a nanotechnology charge. This effort began in 2001, with the creation of the Virginia Nanotechnology Initiative (VNI), designed to foster collaboration among industry, universities, state-based federal laboratories, and relevant state agencies. The initiative was intended to develop partnerships, promote technology transfer, assist in needed workforce development, support some indirect cost recovery on federal research proposals, and commit to nanotechnology advocacy with elected officials. In many respects, the VNI set the stage for 2006 legislation creating the position of secretary of technology for the commonwealth. The secretary and this new department integrated other units across state government and provided new resources, with the charge of "ensuring Virginia remains competitive in cultivating and expanding growth industries, including nanotechnology" (NCSL 2006, 3).

Uncertain Outcomes

One common theme that emerges from the various reports and documents states have generated is an uncritical assertion that new policies and government investment strategies are essential to future development of the nanotechnology sector within their respective boundaries. Such self-assessment is not unique to this area, and indeed, "success stories" tend to supplant other kinds of analysis in governmental efforts to promote economic development (Dewar 1998). Nonetheless, it is striking how little has been written about this activity, other than exceedingly enthusiastic state accounts, whereby multiple states are clearly racing to the top in search of nanotechnology preeminence. As one early scholarly assessment of these efforts notes, "Clearly, the unbridled optimism of nanotechnology advocates needs to be viewed critically. It might take many years before significant economic benefits from nanotechnology are realized, and the ability of states to sequester such benefits within their border is arguable" (Sa et al. 2008, 13). This chapter does not purport to offer a comprehensive analysis, but simply notes that active interstate competition allows for only so many "national leaders" to emerge in any sector, suggesting inevitable disappointments and possible failures as development policies are implemented.

Blending Economic Development and Environmental Protection

Some evidence suggests that economic development opportunities are not the only way states are framing the issue of nanotechnology policy. Several states have begun to consider possible environmental ramifications of nano-technology, primarily focused on possible risks through human exposure, but also somewhat mindful of how nanotechnology might be harnessed to improve environmental quality through various remediation tools that might be developed. Two states, California and Massachusetts, not only match or exceed their counterparts in their willingness to apply economic development strategies to promote nanotechnology, but also give growing indications of considering environmental protection concerns. Both remain at very early stages of this process, not yet approaching the extensive set of policies they have developed in attempting to reduce their greenhouse-gas emissions. Nonetheless, an active process of nanotechnology policy exploration appears to be under way in both Sacramento and Boston. Both states have begun to build on existing policies and institutions and are clearly giving consideration to development of long-term policy options. In turn, neither Minnesota nor Wisconsin has an especially large nanotechnology sector or state development program, but both appear to be following a similar pattern in exploring possible environmental ramifications and policy strategies to govern the use of nanotechnology. Consequently, these cases may offer important insights as to where states may be heading in the future, as well as the possible direction of nanotechnology governance more generally.

California

California's most comprehensive review of nanotechnology policy alternatives is incorporated into a 2005 report by the Blue Ribbon Task Force of Nanotechnology commissioned by State Controller Steve Wesley and U.S. Representative Michael Honda (D-15). This report explores an unusually broad range of issues related to nanotechnology, presented through task forces on commercialization, education, infrastructure and assets, policy and ethics, and research and development. The report makes no bones about California's aspiration to national and global leadership and its perception that nanotechnology represents a profound economic development opportunity for the state. "California is poised to be a world leader in nanotechnology," it proclaims. "Our State has an exceptional existing infrastructure, an unparalleled history of technological success, and an entrepreneurial and innovation culture. Our resources and advantages, as illustrated by our success in semiconductors and biotechnology, are manifold, but we must continue to be proactive in achieving leadership in nanotechnology" (BRTFN 2005, 27).

The report inventories the national nanotechnology landscape and concludes that more than one-quarter of all nanotechnology companies operating in the United States are located in California. It repeatedly invokes the state's prowess in related technological fields and other attributes that position it for an ever larger role as nanotechnology evolves. But woven throughout the report is a cautionary note that other jurisdictions are actively engaged in competition with California, and that the state needs to take continual steps to secure its leadership role. It notes examples such as New York's emphasis on nanotechnology in the Albany area as a possible threat to California and warns that the "competition is intense—from Boston to

Bangalore," underscoring the need for aggressive economic development efforts: "If we do not take action now, another U.S. region or foreign country will assume a leadership role and with it, reap the extensive economic, social and technological benefits accompanying these bourgeoning new industries" (BRTFN 2005, 27). The report also proposes a wide range of economic development strategies to build on established California efforts to promote nanotechnology.

Notably, however, the report moves well beyond economic development to begin to explore the environmental ramifications of nanotechnology. It recognizes the high saliency of environmental issues in California and wants the state to begin to take steps to avoid the controversy that surrounded the adoption of genetically modified organisms in agriculture in many jurisdictions around the world. It envisions lead roles for the California Department of Environmental Protection (CalEPA) and the California Department of Health Services (DHS) in leading "pioneering efforts to understand and communicate the ethical, environmental, and societal implications of nanotechnology." Its proposals are quite general, including "tracking and analyzing emerging health data concerning nanotechnology so that preventative action may be taken before problems develop." These departments are also encouraged to begin "information-exchange" and "problem-sharing" processes with nanotechnology manufacturers to assure "responsible stewardship of nanotechnology products" and also explore ways to inform and educate the citizenry about nanotechnology (BRFTN 2005, 5). "The State of California must address resident concerns about the uses of nanotechnology," notes the report (BRFTN 2005, 10). In response, the California Department of Toxic Substances Control announced in October 2007 that a team of multiple state agencies had begun to "explore ways to minimize environmental and human health risks associated with the manufacture and use of nanotechnology products" (Keiner 2008, 18).

The report integrates economic development and environmental protection by calling on CalEPA, DHS, and other units of state government to take active roles in finding methods to use nanotechnology for environmentally beneficial purposes. It notes considerable potential for nanotechnology to produce new ways to further air and water quality, address groundwater contamination, reduce greenhouse-gas emissions, and promote expanded use of renewable energy. In response, it calls on relevant state government agencies to aggressively explore possible uses of nanotechnology to promote environmentally desired ends, thereby seeking opportunities to become an "early customer" for beneficial applications (BRFTN 2005, 5).

Although the Blue Ribbon Task Force report makes no explicit mention of it, California may already have a regulatory framework of sorts in place that could be directly extended to nanotechnology. In 1986, California voters passed Proposition 65, the Safe Drinking Water and Toxic Enforcement Act, which remains the most far-reaching chemical-disclosure program in the nation (Guber 2003). Prop 65 is somewhat complementary to the national Toxics Release Inventory (TRI), which requires disclosure of the amounts of designated chemicals released to the air, land, and water each year. But it takes a somewhat different approach, including the annual publication of a list of chemicals known to cause cancer or reproductive toxicity. Any firm that releases these chemicals or exceeds what have been established as safe release levels is required to provide a public warning. This is most commonly achieved through a warning label on various products and is determined through a state review process. The proposition is widely seen as having provided not only important information to the public, but also a strong incentive for firms to remove chemicals that would necessitate the application of a warning label to their products (Davies 2007; Graham 2002).

Prop 65 has been in operation for more than two decades and has a well-established set of review and disclosure procedures in place. It currently covers approximately 750 chemicals, although no nanomaterials have yet been added (Keiner 2008). Its extension to nanotechnology in California could be relatively straightforward, just as it has been applied to other new chemicals and products that have emerged since its enactment. Doing so would necessitate an extensive set of studies to determine the environmental health risk of various forms of nanotechnology, and much of this work remains in its very early stages, whether in Sacramento or elsewhere (Laws 2005). Nonetheless, in California, this policy and related environmental disclosure laws are well established.

California further built on its chemical disclosure legacy through 2006 legislation, Assembly Bill 289, which expanded state powers to request information on chemicals' environmental impacts from their manufacturers and importers. The California Department of Toxic Substances Control announced in late 2008 that it would use this new legislation to address nanotechnologies. Earlier that year, Assemblyman Mike Feuer (D-Los Angeles) held stakeholder meetings around the state to discuss the issue and policy options, with expressed intent of developing legislation that would create a nanotechnology regulatory program in California.

Additionally, California has taken the lead nationally in the development of a system for disclosing releases of greenhouse gases. It has secured initial agreement with nearly 40 other states to create a registry for annual releases of carbon dioxide, methane, and other greenhouse gases from a wide range of emissions sources, hoping to use this body of data as a common metric to guide future policy development and implementation. This will build on earlier development of such a registry specifically for California, one of a handful of states to put such a mechanism in place. EPA, meanwhile, has moved much more cautiously into this arena and has explored ways to limit existing TRI policies during the 2000s, thereby ceding greater latitude for leadership to the states (Rabe 2007). EPA accelerated its work on a federal greenhouse-gas inventory in 2009, raising the issue of how a federal program might connect with earlier state and regional efforts.

Consequently, this collective experience could serve to facilitate development of a nanotechnology-focused disclosure and warning program for California, among various policy options. In fact, one California municipality already decided to move in this direction, reflected in the city of Berkeley's 2006 ordinance for "nano-disclosure." Under the California constitution, municipalities retain considerable latitude to tailor their own public health regulations, including those that may exceed state standards. According to the new Berkeley ordinance, "All facilities that manufacture or use manufactured nanoparticles shall submit a separate written disclosure of the current toxicology, to the extent known, and how the facility will safely handle, monitor, contain, dispose, track inventory, prevent release and mitigate such materials." The city seeks toxicity data on inhalation, dermal, oral, mutagenic, and reproductive effects, even though such extensive information simply may not exist in many cases. The ordinance also establishes a strong precautionary emphasis, stating that "if an exposure potential is present but insufficient toxicological information is available, a precautionary approach should be taken which assumes that the material is toxic" (City of Berkeley 2006). Berkeley is not a particularly large city, with a population of less than 110,000 in the 2000 U.S. Census, but it is home to a considerable research infrastructure with nanotechnology interests, including units at the University of California's Berkeley campus that have raised concerns about the possibility of research constraints that might be imposed by the city ordinance. This step sets in place the most restrictive regulatory program for nanotechnology of any government in the United States, and it is lodged in a state known for taking aggressive environmental protection action of its own. According to Berkeley mayor Tom Bates, "I would love for us to continue to be on the forefront and continue to put forward new and innovative ideas that allow it [nanotechnology development] to happen but does so in a way that makes sure the public is safely protected" (Keiner 2008, 7).

Massachusetts

If any state rivals California for national leadership in nanotechnology development and in beginning to consider environmental protection ramifications, it is most likely Massachusetts. The commonwealth already has a sizable body of nanotechnology operations and has hoped to build on its large base of research universities to further expand this role. It has explored a number of mechanisms to promote this area of economic development, led by the Massachusetts Technology Collaborative (MTC), a quasipublic entity that provides support for a range of emerging technologies, nanotechnology included. The collaborative is part of a massive state effort to promote economic growth through investments in new technologies, led in many ways by its substantial life-sciences sector. In 2004, the MTC issued a major report with recommendations to further develop the nanotechnology sector in Massachusetts (MTC 2004).

Like California, Massachusetts has begun to give thought to environmental protection considerations. The state has a long record of regional and national leadership in various areas of environmental affairs, and it appears to be building a model for thinking through nanotechnology policy options based on its experience with drinking-water regulation earlier in the decade. In that instance, concerns about perchlorate contamination in drinking water on Cape Cod led to the formation of a working group that cut across traditional departmental boundaries and resulted in the creation of a new statewide standard in 2006.

This experience fostered the formation of an Emerging Contaminants Workgroup in 2007 to attempt to identify new environmental challenges and begin to determine whether some form of policy response was warranted. Nanotechnology has been designated as one of the six initial priorities of this workgroup, which includes representatives from the Massachusetts Department of Environmental Protection (DEP), Department of Public Health, Office of Technical Assistance, and Water Resources Authority. The workgroup also actively involves the Toxics Use Reduction Institute, based at the University of Massachusetts at Lowell, which is authorized by state law to support implementation of the state's Toxics Use Reduction Act

(TURA). A senior official from the state DEP noted in late 2007 that they were just getting started, but the official believed that the multiunit network mechanism increased the likelihood for a broad, anticipatory approach to any environmental challenges nanotechnology might pose. According to one early analysis, the workgroup centralizes the DEP's "focus on emerging contaminants, fosters information exchange, and brings together a broad range of cross-program expertise" (McCarty and Bosso 2008, 21).

Much like California, Massachusetts also has considerable expertise that might be parlayed into early engagement on nanotechnology. The common-wealth has long been active in issues of toxics contamination and reduction, reflected in part through TURA and its long-standing involvement on pollution prevention issues. The city of Cambridge also plays host to major research universities and has a history of early involvement in a number of technology issues, including a 1977 ordinance on recombinant DNA. Cambridge environmental health officials have been in conversation with their Berkeley counterparts and attempted to engage diverse stakeholders in considering possible options but, as noted in Chapter 4, have thus far decided against any formal policy steps.

Minnesota and Wisconsin

The experiences of two midwestern states with long traditions of innovation in environmental policy further confirm the possible expansion from the Massachusetts case in using an "emerging contaminants process" to highlight the issue of nanotechnology and begin to formulate policy options. Each of these states is developing its own distinctive processes, yet there are obvious points of overlap between them. Indeed, leaders of these efforts in both states are clearly aware of the work of their neighboring state and appear to be engaged in an active process of policy learning, with ideas and even exact phraseology flowing back and forth across the Mississippi River dividing Minnesota and Wisconsin. Neither state ranks among national leaders in the current scale of nanotechnology involvement, but both have some degree of related activity under way within their borders.

Minnesota has a long-standing record of attempting to anticipate emerging environmental problems and fashion strategies designed to minimize potential risks. This includes very early engagement in pollution prevention, strategies to foster coordination across traditional medium divides such as air and water, and efforts to reduce releases of toxic chemicals (Rabe 1999). On occasion, this has led to far-reaching experiments in regulatory innovation, offering possible lessons for other states and even the nation (Marcus et al. 2002). All of this prior experience was evident in 2006, when the Minnesota Pollution Control Agency created an Emerging Issues Team (EIT). Somewhat similarly to the Massachusetts process, this involved the formation of a team of environmental policy professionals to attempt to examine a range of issues that were starting to surface as environmental concerns and begin to formulate policy options. One key mission of the EIT is to explore "new areas of environmental concern that are not currently incorporated into regular environmental protection activities in Minnesota" (Crane 2007). As one team member noted, nanotechnology is high on their list, and it has indeed received considerable attention in the early work of the EIT.

EIT members have expressed particular concern over the absence of any standard or reliable method to measure nanoparticles, much less consider their potential risk. They highlighted possible human health hazards, ranging from lung damage in lab rats posed by carbon nanotubes to damage to human DNA, including possible cancer risk, from manufactured nanoparticles. At the same time, the EIT is mindful of potential environmental benefits that could be derived from expanded development of nanotechnologies, including waste reduction, cleanup of industrial contamination, drinking-water protection, and improved energy production and consumption. Relevant technologies might include sensors to create more precise measurement of toxic chemicals and viruses and remediation strategies using nanosize rust particles to remove arsenic from drinking water (Crane 2007). These possible environmental "goods" that might emerge from nanotechnology were not discovered by the EIT but have clearly emerged as a major focal point in its deliberations and build on a Minnesota tradition of advancing new technologies and ideas to promote environmental protection.

Wisconsin may be somewhat farther along in its policy development process, owing to a 2006 report prepared for the Wisconsin Department of Natural Resources by its Nanotechnology and Natural Resources Task Group. Known as the Nanotechnology White Paper, this report was developed by a staff team headed by an official of the Integrated Science Services Bureau, with members drawn from the Bureaus of Remediation and Redevelopment, Air Management, Fish Management, Waste and Materials Management, and Cooperative Environmental Assistance. The team was created in order "to identify and describe the issues facing the Department with regards to nanotechnology" (Griffin et al. 2006). The report presents an unusually broad range of concerns and policy options and could indeed set the agenda for an extensive policy response. As one Minnesota official has noted, "As a state agency, the Wisconsin DNR has probably done the most work in considering how they would deal with nanotechnology in their regulatory programs and in risk assessments" (Crane 2007).

The task group is clearly aware of existing data gaps, including point emissions of nanomaterials into air and water, nonpoint air and water sources of nanomaterials, amounts and types of nanomaterials in waste-management facilities, and accidental releases, among others (Powell et al. 2008). Its members are also cognizant of potential environmental applications of nanotechnology, much like their colleagues in Minnesota, placing particular emphasis on remediation and pollution control, wastewater treatment, pollution detection and environmental monitoring, and "green" manufacturing. According to the report, the team believes that "nanotechnology shows enormous promise for environmental cleanups" (Griffin et al. 2006, *17*). But perhaps most significant is an uncommonly thorough review of ways that virtually all existing Wisconsin environmental legislation might eventually apply to nanotechnology. The report also considers existing state innovations and experimental programs that might prove relevant, such as the FACT System program, which provides citizens access to substantial information on emissions and discharges, and the Green Tier program, which combines beyond-compliance opportunities with more efficient permit approval.

In essence, the task group report represents an unusually comprehensive overview of Wisconsin environmental governance. It considers virtually every conceivable way nanotechnology might need to be considered from an environmental protection perspective or how it could foster integration across traditional program boundaries. The report also goes farther than any other state analysis in outlining four broad options for state governance of nanotechnology concerns: a cross-cutting Nanotechnology and Natural Resources Team; a single point of contact (SPOC), who would emerge as the departmental lead staff person on all matters relevant to nanotechnology; an ad hoc scientific review committee; or the current status quo, wherein nanotechnology does not have a formal place. The report endorses the SPOC approach as the best option for Wisconsin and also outlines the resources that would be necessary to assure its implementation if adopted. It readily acknowledges the uncertainty of resource allocation for such an undertaking, the complexity of the cross-program collaboration, and the possible risk to nanotechnology development in Wisconsin as a result of the "potential external perception of the department overreaching its authority" and the "potential perception of inhibiting industry innovations with added regulations" (Griffin et al. 2006, 29).

Nonetheless, the report consistently emphasizes that nanotechnology presents a unique set of opportunities for both environmental protection and economic development in the state and concludes:

> It is imperative that the Department's approach to addressing nanotechnology and its implications for environmental protection be pro-active. The effects of a pro-active approach versus a reactive approach will encourage new ways for the Department to tap into the benefits of this technology without ignoring the risk; and at the same time allow flexibility to continue to address new issues as nanotechnology evolves and expands. (Griffin et al. 2006, 30)

The DNR leadership has not formally responded to the report, although one state representative has announced plans to introduce legislation that would establish a nanotechnology reporting system or registry.

LOOKING AHEAD

There is little evidence that any large set of states is racing to the top to erect demanding environmental protection programs for nanotechnology. Unlike with climate change and other areas of environmental concern, state policy roles are modest thus far and seem focused overwhelmingly on advancing economic development opportunities. Exceptions to this pattern have been emerging, such as California and Massachusetts, and the number of states moving in this direction may increase should the nanotechnology issue gain saliency and public concerns over environmental safety mount. But just as national institutions remain in the very early stages of thinking about possible policy interventions, few lessons can be drawn thus far from evolving state experience. As Leslie Carothers of the Environmental Law Institute has noted, "Much more attention needs to be paid to law and governance policy as the scientific assessment moves forward" (Carothers 2006, 64).

One early lesson is that no one governmental unit or policy approach is likely to address all issues related to nanotechnology in a comprehensive way. At the federal level, it appears inevitable that jurisdiction will be shared through some form of interunit collaboration among entities such as the Environmental Protection Agency, Food and Drug Administration, Occupational Safety and Health Administration, Consumer Product Safety

Commission, Department of Agriculture, Department of Energy, National Institute for Science and Technology, and U.S. Patent and Trade Office, among many others. Similarly, at the state level, California has begun to turn to multiple departments, just as Massachusetts and Wisconsin envision some form of working-group approach across diverse units of state government.

States will need to find ways not only to bring together players that can guide environmental protection considerations, but also to link these with the active set of institutions focusing exclusively on nanotechnology growth and development. This is likely to necessitate a type of governance that defies conventional hierarchical patterns. Indeed, all of the skills embedded in network approaches to governance will likely prove essential (Goldsmith and Kettl 2009). This includes finding ways to break through traditional departmental and agency rivalries to secure collaboration on the complex issues that will be inherent in any future nanotechnology policy, whether it involves disclosure or regulation, or takes some other form.

Models for such networking exist in environmental protection, including the multiunit framework developed to guide biotechnology policy in the 1980s and 1990s (Elliott 2005; Kuzma 2006). At the state level, some recent climate policies have demonstrated considerable ability to bridge traditional divides among units responsible for environmental protection, energy, transportation, and other areas and begin to forge new approaches that may have considerable potential to reduce greenhouse-gas emissions and facilitate some degree of in-state economic development. These offer important lessons for a federal government that is increasing its focus on this issue. The most recent climate-change developments at the state level involve formal collaboration among neighboring states, most notably the efforts of 23 states and 4 Canadian provinces located on both coasts and in the Midwest to create regional carbon cap-and-trade systems. There is no comparable indication of multistate collaboration related to nano-technology, particularly given the prevailing emphasis on intrastate economic development. But governance networks take many forms, and the state government role in nanotechnology may evolve in ways that are difficult to envision at present. As a report from the International Risk Governance Council has noted, "The question is really how to move beyond simplistic notions, such as self-regulation, to building systems of accountability and governance that are conducive to appropriate expansion of both science and democracy" (IRGC 2005, 119).

In some respects, the area of nanotechnology policy appears to be following an intergovernmental path similar to that of climate change.

In both instances, prolonged inertia among federal institutions created a window of opportunity for states to begin to explore innovative strategies. A steadily expanding number of states over the past decade have developed "homegrown" policy approaches, many focused on economic development opportunities via early action. A growing number of them, however, are also explicit about using these policies to reduce their greenhouse-gas emissions. Such bottom-up cases provide tremendous opportunity for policy learning to occur across jurisdictional lines. This is clearly under way already between Minnesota and Wisconsin, and it could expand to other states and regions, and even federal policy.

Moreover, as states gain growing experience in this area, they can help shape and define what might be feasible and effective for other regions or on a national basis. In addition, they will likely play a central role in implementation of any future policy, as has been the case in medium-based pollution programs, so their growing awareness and expertise could prove invaluable. Intergovernmental transfer of models and lessons is never easy, especially in the American context. But it underscores the importance of early state policy initiative and at least the possibility of intergovernmental policy development based on real expertise and best practice, rather than whatever proves popular in Washington, in the event that the policy agenda opens to allow consideration of a particular issue.

Note: I am grateful to Christopher Bosso, J. Clarence Davies, Patrick Hamlett, Jennifer Kuzma, Michael Mintrom and external reviewers for thoughtful suggestions on earlier versions of this chapter. An earlier version was presented at the 2007 Annual Meeting of the American Political Science Association.

REFERENCES

BRTFN (Blue Ribbon Task Force on Nanotechnology). 2005. *Thinking Big about Thinking Small: An Action Agenda for California.* Sacramento, CA: BRTFN.

Carothers, Leslie. 2006. Governance Structure for Nanotechnology No Small Task. *Environmental Forum* (January–February): 64.

City of Berkeley. 2006. Municipal Code Section 15.12.040 on Manufactured Nanoparticle Health and Safety Disclosure. City of Berkeley, California. www.ci.berkeley.ca.us/citycouncil/2006citycouncil/packet/121206/2006-12-12%20Item%2003%20-%20Ord%20-%20Nanoparticles.pdf (accessed March 12, 2009).

Crane, Judy L. 2007. State Regulatory Perspectives on Nanotech Oversight. Slides from October 4 presentation. St. Paul, MN: Minnesota Pollution Control Agency.

Davies, J. Clarence. 2006. *Managing the Effects of Nanotechnology.* Washington, DC: Woodrow Wilson International Center for Scholars.

——. 2007. *EPA and Nanotechnology: Oversight for the 21st Century.* Washington, DC: Woodrow Wilson International Center for Scholars.

Davies, J. Clarence, and David Rejeski. 2007. Overseeing the Unseeable. *Environmental Forum* (November–December): 36–40.

Dewar, Margaret E. 1998. Why State and Local Economic Development Programs Cause So Little Economic Development. *Economic Development Quarterly* 12 (1): 68–87.

Derthick, Martha. 2007. Going Federal: The Launch of Medicare Part D Compared to SSI. *Publius: The Journal of Federalism* 37 (Summer): 351–70.

Elliott, E. Donald. 2005. Regulate Nano Now. *Environmental Forum* (July–August): 43.

Fiorino, Daniel. 2006. *The New Environmental Regulation.* Cambridge, MA: MIT Press.

Goldsmith, Stephen, and Donald F. Kettl, eds. 2009. *Unlocking the Power of Networks: Keys to High-Performance Government.* Washington, DC: Brookings Institution.

Graham, Mary. 2002. *Democracy by Deliberation.* Washington, DC: Brookings Institution.

Griffin, M.P., G. Edelstein, J. Myers, C. Schrank, L. Sukup, and G. Wheat. 2006. *Nanotechnology and Natural Resources: Preparing the Department for the Present and the Future.* Madison, WI: Wisconsin Department of Natural Resources.

Guber, Deborah Lynn. 2003. *The Grassroots of a Green Revolution: Polling America on the Environment.* Cambridge, MA: MIT Press.

IRGC (International Risk Governance Council). 2005. *Survey on Nanotechnology Governance. Vol. A: The Role of Government.* Geneva, Switzerland: IRGC.

Jones, Charles O. 1975. *Clear Air: The Policies and Politics of Pollution Control.* Pittsburgh, PA: University of Pittsburgh Press.

Keiner, Suellen. 2008. *Room at the Bottom? Potential State Strategies for Managing the Risks and Benefits of Nanotechnology.* Washington, DC: Woodrow Wilson International Center for Scholars.

Kettl, Donald F. 2007. *System under Stress: Homeland Security and American Politics,* 2nd ed. Washington, DC: CQ Press.

Kraft, Michael E. 2010. Environmental Policy in Congress. In *Environmental Policy: New Directions for the Twenty-First Century,* 7th ed, edited by Norman J. Vig and Michael E. Kraft. Washington, DC: CQ Press, 99–124.

Kuzma, Jennifer. 2006. Nanotechnology Oversight and Regulation—Just Do It. *Environmental Law Review* (December): 10913–23.

Laws, Elliott P. 2005. Unreasonable: Nano Risks Still Unknown. *Environmental Forum* (July–August): 12.

Marcus, Alfred. A., Donald A. Geffen, and Ken Sexton. 2002. *Reinventing Environmental Regulation: Lessons from Project XL.* Washington, DC: Resources for the Future.

McCarty, Katrina L., and Christopher J. Bosso. 2008. Case Study: The Massachusetts Department of Environmental Protection and Environmental Regulation Amendments for Biotechnology. Unpublished manuscript of the Nanotechnology & Society Research Group, Northeastern University, Boston.

MTC (Massachusetts Technology Collaborative and Nano Science and Technology Institute). 2004. *Nanotechnology in Massachusetts: A Report on Nano-Scale Research and Development and Its Implications for the Massachusetts Economy.* Boston, MA: Massachusetts Technology Cooperative.

Mintrom, Michael. 2009. Competitive Federalism and the Governance of Controversial Science, *Publius: The Journal of Federalism* 39 (4) (Fall): 606-631.

Nakamura, Robert T., and Thomas W. Church. 2003. *Taming Regulation: Superfund and the Challenge of Regulatory Reform.* Washington, DC: Brookings Institution.

NCSL (National Conference of State Legislatures). 2006. *Nanotechnology Statutes.* Washington, DC: NCSL.

NNI (National Nanotechnology Initiative) Workshop. 2003. *Regional, State, and Local Initiatives in Nanotechnology.* Washington, DC: Nanoscale Science, Engineering, and Technology Subcommittee.

Powell, Maria, Martin Griffin, and Stephanie Tai. 2008. Bottom-Up Risk Regulation? How Nanotechnology Risk Gaps Challenge U.S. Federal and State Environmental Agencies. *Environmental Management* 42 (3): 426–43.

Rabe, Barry G. 1999. Federalism and Entrepreneurship: Explaining American and Canadian Innovation in Pollution Prevention and Regulatory Integration, *Policy Studies Journal* 27 (2): 288–306.

———. 2004. *Statehouse and Greenhouse: The Emerging Politics of American Climate Change Policy.* Washington, DC: Brookings Institution.

———. 2007. Environmental Policy and the Bush Era: The Collision between the Administrative Presidency and State Experimentation. *Publius: The Journal of Federalism* 37 (Summer): 413–31.

———. 2008. States on Steroids: The Intergovernmental Odyssey of American Climate Policy. *Review of Policy Research* 25 (March): 105–28.

Sa, Creso M., Roger L. Geiger, and Paul M. Hallacher. 2008. Universities and State Policy Formation: Rationalizing a Nanotechnology Strategy in Pennsylvania, *Review of Policy Research* 25 (January): 3–19.

Scheberle, Denise. 2004. *Federalism and Environmental Policy: Trust and the Politics of Implementation,* rev. ed. Washington, DC: Georgetown University Press.

Smith, David M., and Christopher J. Bosso. 2007. *Realizing the Potential of the Massachusetts Nanotechnology Sector: Recommendations for the Commonwealth.* Report of the Nanotechnology & Society Research Group, Northeastern University, Boston, MA.

Teske, Paul. 2004. *Regulation and the States.* Washington, DC: Brookings Institution.

CHAPTER 7

NANOTECHNOLOGY AND TWENTY-FIRST-CENTURY GOVERNANCE

Christopher J. Bosso and W. D. Kay

This first decade of the twenty-first century finds most U.S. government institutions, regulatory agencies in particular, facing a rather serious dilemma. On the one hand, to use the now fashionable neologism, they are "under-resourced." As the other chapters in this volume have discussed, environmental regulators at all levels are hard-pressed to meet current obligations, let alone deal with the potential fallout from nanotechnology's promised "next Industrial Revolution." And, not to put too fine a point on it, the fundamental problems to which our authors have referred throughout have been exacerbated by nearly three decades of often bipartisan antigovernment rhetoric, accompanied by mandated regulatory devolution, if not complete deregulation, all of which raised hard questions about the effectiveness of the U.S. regulatory regime even before the financial market and regulatory crises of 2008 (Frederickson and Frederickson 2006).

To makes matters worse, it will not be enough for a new administration and Congress to fix or restore eroded regulatory capacity by simply boosting agency budgets. The deeper problem, as foreshadowed in Chapter 1, is a fundamental matter of institutional capacity and the expected informational and analytical needs of agencies and officials that researchers, firms, investors, and citizens expect to make critical decisions on a wide range of manufactured nanomaterials and applications. In this regard, if even a portion of nanotechnology's promise comes to pass, this will amount to a wave of materials and applications for which we are just not ready on any number of societal, ethical, and policy dimensions.

LOOKING BACK TO THE FUTURE

One irony is that we have, in a sense, been here before. A bit less than a century ago, scientists all over the world were being seized with similar revolutionary fervor, inspired by a rapid series of exciting developments in the then newly emerging field of nuclear physics (Kay and Eijmberts 2009). Called by some the "golden age of physics" (Weisskopf 1970), the first three decades of the twentieth century saw the discovery of the atomic nucleus, the neutron, artificial radioactivity, and nuclear fission, to name but a few scientific breakthroughs. And it was not basic matter or new materials that motivated these researchers or those who followed them, but rather the energy that might be obtained from the atomic nucleus. In 1903, the chemist Frederick Soddy, Ernest Rutherford's colleague at McGill University and later a Nobel laureate for his research on radioactivity, told an audience that a pint bottle of uranium contained enough energy to propel an ocean liner from London halfway across the globe to Sydney and back. In his 1908 book, *The Interpretation of Radium*, Soddy envisioned power that "could transform a desert continent, thaw the frozen poles, and make the whole world one smiling Garden of Eden" (Weart 1988, *x*). In 1923, a then unknown Enrico Fermi wrote that a gram of matter equaled more than the total energy of a 1,000-horsepower motor for three years (Badash et al. 1986). Compare these predictions with President Bill Clinton's exuberant claim in 2000, on ushering in the National Nanotechnology Initiative, that nanotechnology will lead to advances in computing power that will make it possible to shrink "all of the information housed at the Library of Congress into the device the size of a sugar cube" (White House 2000), and one gets the feeling that, if not promising a "new Industrial Revolution," the pioneers of nuclear researchers did not lack for vision.

The "golden age of physics" also influenced other scientific fields in much the same way that nanotechnology promises to poke holes through all kinds of disciplinary walls and reorient much of basic and applied science and engineering. For example, Max Delbrück, a colleague of Lise

Meitner (codiscoverer of nuclear fusion), is better known today as one of the founders of molecular biology. Completely new subfields such as X-ray and gamma-ray astronomy emerged in such disciplines as astronomy and astrophysics. The understanding of nuclear decay led directly to the carbon-dating techniques widely employed in geology and paleontology. So if nanoscience is often depicted as an enabling scientific platform, it is certainly not the first.

Yet there is a difference—so far. The status of nuclear science as a technology enabler is obscured by the deep controversies that surround its two most well-known—some might say infamous—applications, namely weaponry and power generation. The historical irony is that these were by no means its only uses, nor were they even the first. During the early 1900s, Marie Curie used bits of pure radium, applied directly to tumors, as a type of cancer treatment (Badash et al. 1986). With the discovery of artificial radioactivity, it became possible to create a wide variety of radioisotopes, opening up completely new approaches to medical diagnosis and treatment. Shortly after World War II, the General Atomics Corporation began producing the TRIGA (Training, Research, Isotopes, General Atomic) reactor for use in universities and laboratories. In addition to becoming the best-selling nuclear reactor ever made, TRIGA was, and remains, the only such device to turn a profit (Dyson 2002). Nuclear propulsion dates back to 1955, with the launching of the *Nautilus* submarine. Surface ships have also been nuclear-powered, including an icebreaker manufactured in Russia.

Many in the early nuclear science community harbored more grandiose plans. Up through the end of the 1960s, serious design work was begun (but eventually dropped) on a nuclear-powered aircraft—it would only have to land to let the crew reenlist, went the joke at the time—a nuclear locomotive, and two types of nuclear-powered spacecraft, one of which, Project Orion, was to have been propelled by a series of atomic explosions. In 1957, the Ford Motor Company unveiled plans for the Nucleon, which envisioned a mini-reactor in the trunk of a car and was projected to take a driver 5,000 miles on a single charge. In the end, the company produced only a $^3/_8$-scale unpowered model.

If many of these plans foundered because of technical or economic hurdles, other proposed applications proved problematic because of deep-seated public unease about nuclear power in the wake of its spectacular and destructive wartime debut. For example, the irradiation of food using very low-level gamma rays can destroy harmful bacteria and preserve freshness for extended periods. The technique is used in more than 40 countries and is especially useful where refrigeration is not readily available, but food producers in developed nations in particular have been hesitant to adopt this technology, fearing consumer rebellion about "nuking" their milk or meat (Meins 2003). At the same time, however, radiation-based tests are used to detect stresses and metal fatigue in such structures as bridges and overpasses. In sum, public receptivity to the fruits of nuclear science is greatly contextual, driven by applications and framed through a path-dependent social construct—starting with the nuclear bomb and reinforced by Three Mile Island and Chernobyl—that to a great extent has limited its utility in everyday life.

The lessons for nanotechnology in this story seem clear. An enabling technology looks particularly promising and spawns widespread prognostications about its transformative—even revolutionary—impacts, but some applications are constrained or rejected outright because of public anxiety about potential risk. Government and scientific elites are first reflexively boosterish, proclaiming the great and beneficial changes to come (Berube 2006), and then concerned when some uses or mishaps spark public debate and resistance, thereby limiting (they fear) the technology's potential— what has been labeled the "wow to yuck trajectory" (Kulinowski 2004). The revolution, although not derailed, is nevertheless constrained, leaving behind disillusionment among the faithful that the public is mired in irrational fears about technology.

New technologies, whether nuclear fission, synthetic chemical pesticides, or genetically modified organisms, eventually pose a range of environmental and health problems, some unique to that technology and others of a generic kind. Moreover, technology boosters and free-market advocates have always been slow to admit a need for some anticipatory government attention to these potential effects—even when they are strongly suspected—with subsequent impacts on available resources and room to maneuver for those who are eventually given the task of responding to now apparent side effects. From this perspective, one might even argue that we are rushing ahead with nanotechnology development and commercialization far more quickly than with nuclear fission. In this sense, it is not too soon to ask how we might handle the potential side effects of nanotechnology even before they are made manifest. We have been down this road before.

LEARNING TO LIKE UNCERTAINTY?

> Information is the foundation for the twenty-first century. Trust is key to useful information. And trust has long proven elusive in environmental politics. (Kettl 2002, *185*)

The first question driving this overall project focuses on how government deals with conditions of profound uncertainty about the potential environmental and health risks of an emerging technology. Marc Eisner pointed to this essential dilemma in environmental regulation when he argued in Chapter 3 that uncertainty erodes the capacity of regulators to make choices, often leading them to fall back on potentially outmoded or misdirected procedural certainty—hard rules and procedures, as it were—when greater nuance may be called for. For example, little agreement exists as to whether the Toxic Substances Control Act has proven effective in reducing uncertainty about toxicological risk, yet 30 years after enactment, it stands as the U.S. government's primary tool in dealing with thousands of chemicals. We have clearly laid-out processes for registering products, yet very little real knowledge about them or the relative risks they may pose. This dilemma is made more acute because TSCA is likely to be the federal government's primary tool in regulating nanoparticles—about which even less is currently known. As the noted environmental policy expert J. Clarence Davies testified before a House subcommittee addressing changes in the law, "TSCA is not serving us well now and it will not in the future" (Davies 2009).

An obvious solution to acute uncertainty is to lessen it through more research, to obtain greater precision about risk profiles through more and better information. And in this regard, the federal government in particular is beginning to invest significant funds for basic research in agencies like the National Institute of Occupational Safety and Health and through university-based initiatives such as the Centers for Environmental Implications of Nanotechnology at UCLA and Duke, recipients of a five-year $25 million joint National Science Foundation–Environmental Protection Agency grant to generate new scientific knowledge about the toxicity of nanomaterials. Such research will yield more and better data on everything from exposure levels to workers and other affected populations to the broader environmental impacts of nanomaterials throughout their life cycles.

But as political scientist Don Kettl reminds us in thinking about the next generation of environmental policy, a dilemma is inherent in information: "It is everything, and is nothing" (Kettl 2002, *183*). Even with more and better information, what would we do with it? Despite considerable effort to construct ever more precise decision models to aid in making choices, information by itself does not—and should not—decide. To regulate is to make choices, and to choose involves values, a central one being the degree to which the broad range of policy stakeholders—citizens especially—are willing to embrace some degree of uncertainty about risk in return for other benefits. The desire to want to know and, thus armed, to be prepared runs up against a reality that the sobering complexity posed by nanotechnology, in and of itself or as an enabling technology, erodes that capacity to know for everyone, even as new products and applications flood into everyday use. Thus is the inherent tension within the notion of precaution: our desire to be certain confronts the reluctant realization that we cannot. We face an inevitable trade-off between being anticipatory and having sufficient knowledge. The more anticipatory we try to be, the less information is available (even if it could be collected in a timely way) on nanomaterials and their impacts.

This does not mean that we should proceed blithely down the path of acceptance. Asking citizens in particular to live with some degree of uncertainty about risk will require far more transparency on the part of regulators and regulated industries alike. A greater societal willingness to admit that we can never reduce risk to zero cannot be attained without a far more open and honest debate about proportional and acceptable risk—and for whom. Marc Landy's notion, outlined in Chapter 5, of a "grand bargain" needed to move to a "weight of evidence approach" goes far beyond his focus on EPA. It requires the building of greater capacity of and more trust in the institutions of government than currently exist. It also requires the development of a fair amount of trust in those who govern. In the end, dealing with uncertainty about risk is not a technical or informational problem, but one of democratic governance. We return to this theme later.

THE ROLE OF THE REGULATED

The second question that guided this study was, given such uncertainty about nanotechnology's health and environmental effects, as well as the information asymmetries that obtain in such instances, how does government structure its relationship with the regulated? The core chapters in this book each in some way or another have spoken to this critical relationship. Marc Eisner focused on the essential dilemma of regulating that which we may not completely understand, which inevitably forces government regulators to accept some degree of discretion to business. As he argued, however, discretion—even in its most robust form, as self-regulation—is not just a synonym for deregulation. Rather, he sees a regime based on "regulatory pluralism," with a fair amount of industry self-regulation to provide for more flexibility managing risks combined with government-mandated oversight and, possibly, industry-wide controls once certain risks become clearer and proportionally situated.

Cary Coglianese further explored this fundamental relationship between regulators and business in Chapter 4 and concluded that the dilemma Eisner posed may never be entirely resolved. He argued that granting business significant regulatory discretion can be achieved only within a broader oversight system. Yet the very reasons that make delegating discretion to industry seem attractive, even necessary, in the current context of nanotechnology development are reasons to suspect the effectiveness of voluntary or discretionary efforts. After all, nanotechnology is not a single sector like nuclear power or governed through a single trade association—regardless of efforts in the United States and European Union to construct one. Indeed, "nanotechnology" as a noun is fast being transformed into "nano" as adjective—nanomedicine, nanoelectronics, nanoenergy, nanoagriculture, and so forth—with consequent fragmentation of industry needs and regulatory challenges into communities defined by specific uses (e.g., consumer products, medical applications), potential side effects (e.g., nanoparticle toxicity), or affected populations (e.g., the poor). In Coglianese's terms, there will be no single peak industry association to deal with, nor any single set of international standards to guide, an environmental management system. Then what?

At its nub, this is a broader debate about what political arrangements are necessary to foster appropriate behavior among the regulated. No procedural tinkering between regulatory agencies and affected business sectors will ultimately substitute for more and better representation of affected stakeholders, particularly citizens, to balance business interests in regulatory decisionmaking (Beierle and Cayford 2002). As Coglianese has argued elsewhere, however, even with more and readily accessible information, the sheer complexity of the regulatory process undermines the capacity of average citizens to participate on anywhere close to an even level with business (Coglianese 2007a). Yet if such complexity undermines the kinds of direct, broad-based citizen participation sought by proponents of "strong democracy," fostering a more robust pluralism of informed representation is not so far-fetched. In some ways, it already exists in the current community of major national environmental groups, long-standing organizations such as the Environmental Defense Fund and Natural Resources Defense Council, which have the scientific and legal capacity as well as the political clout to act as representatives of the broader public good within a more transparent informational milieu (Bosso and Rodrigues 2007).

Both Coglianese and Barry Rabe, in Chapter 6, also pointed to the case of Cambridge, Massachusetts, where an appointed citizens advisory committee deliberated about how the city should approach the possible regulation of nanomaterials produced or manipulated within city boundaries. Likely keys to public acceptance of the committee's decision *not* to impose any regulations at this time were its diverse and expert composition, including academic toxicologists, lawyers, and public health officials; its legitimacy with the residents of Cambridge; and the opportunities the process afforded for citizen input. Again, the crucial point here is that the committee contained the requisite expertise to assess information and enjoyed the trust of local residents to make reasoned judgments about risk. That Cambridge in the previous 30 years had developed the requisite experience with another emerging technology, recombinant DNA, certainly helped, if only because it gave residents faith that city officials were asking the right questions (Lipson 2003).

If, as Eisner and Coglianese both argue, the current absence of certainty about health risks posed by nanomaterials combined with the asymmetries in information possessed by business compel government to afford greater discretion to the regulated, it should also compel greater formal representation of nonbusiness interests in decisionmaking. In a variation of Landy's "grand bargain," a regime that leans toward informed self-regulation should come accompanied by more expansive and formal inclusion of advocacy groups, science advisory panels, and expert citizens in oversight and regulatory deliberations. It will require expanding the narrowly configured relationship between regulators and the regulated into a broader discursive network of interests. In short, it will require more, not less, overall participation in deliberations about risk, the absence of which at the height of the Cold War arguably shaped citizen resistance to the development of nuclear energy. To be sure, the cost of greater overall participation is a more fractious discourse. The cost of *not* doing so is public suspicion of the validity of the decisionmaking process and, by extension, of the technology in question. Indeed, one can argue that pan-European hostility to genetically modified crops had far less to do with technology per se than with deep-seated and historically rooted skepticism about the honesty, efficacy, and transparency of business and government elites.

THE QUESTION OF AGENCY CAPACITY

Achieving a more robust pluralism in regulatory oversight and decisionmaking under conditions of uncertainty requires greater government capacity to ultimately act as arbiter of conflicts and endow any outcomes with necessary democratic legitimacy. Indeed, the larger project that animates this study is ultimately about the capacity of government in the twenty-first century. "Capacity" is a concept that is widely employed in many literatures but, surprisingly, seldom examined closely. For decades, scholars in fields ranging from political development to regulation, not to mention the life sciences and engineering, have identified capacity-building in one form or another as an absolute prerequisite for achieving stated ends or missions. A large body of literature, particularly in public policy studies, also portrays the dire consequences of its *absence*: lacking capacity, or having insufficient capacity, is frequently cited as a cause of failure at all levels of government and, on occasion, of government itself.

But what *is* capacity, exactly? What do we really mean in bandying this term about, often casually, in thinking about the purpose of government? Most frequently, we think, it is taken simply to refer to the means by which a desired goal is attained. Unfortunately, given the wide usage referred to above, this could mean almost anything. Thus it is hardly surprising that the literature on capacity and capacity-building, such as it is, is highly fragmented. For some authors, "capacity" refers narrowly to resources; to others, it is a function of management; still others see it as an organizational or institutional feature; and so on.

There is good reason to believe that a more generalized and overarching notion of capacity, and of capacity-building, is needed. Every so often—and such occasions are sure to be more frequent and challenging in the years ahead—governments need to, in essence, retool. Sometimes this is a relatively straightforward process, as when public officials respond to local population growth by increasing the number of police units or firefighters—a matter of simple addition. Most often these additive decisions are guided by tested formulas or analogies to similar organizations or government units. Other cases, unfortunately for decisionmakers, are far trickier. For example, developing nuclear power as a weapon and a source of energy required government organizations—initially the U.S. Army and later the Atomic Energy Commission—to take on a task that no government had ever performed before.

As the nuclear power example suggests, emerging technologies, including and not limited to nanotechnology, pose real challenges to government capacity. These developments will—and in some areas, already do—place considerable strain on a range of federal agencies, including the U.S. Patent and Trade Office, National Institute of Standards and Testing, Environmental Protection Agency, and Food and Drug Administration, to name only a few. Most, if not all, of these agencies are already overburdened by present workloads and by legacies of operations that may have less relevance to new challenges. As a result, mandates to deal with waves of completely new and often not fully understood technologies should raise major concerns for those who worry about the ability of democratic government to foster technological innovation while also protecting the public health and welfare against the inevitable side effects of any new technology.

But technical change is only part—albeit a large part—of the problem. Increasing globalization, with consequently rapid expansion in the diversity of markets and service populations, will inevitably lead to new and unique demands on public institutions. So, too, will the ever-shifting requirements of national and homeland security. In short, it is no longer sufficient to regard government capacity-building in an ad hoc, as-needed fashion. Public organizations, in the near future, will be called on to develop new skills and capabilities—in effect, to upgrade—on a sustained basis.

By "unpacking" the concept of capacity, we hope to learn how to build it more readily and avoid the "reinventing the wheel" syndrome that characterized earlier efforts. In this regard, much writing vastly oversimplifies the concepts of capacity and capacity-building, at least implicitly. Authors have a marked tendency, for example, to regard capacity as an attribute of single entities, such as (depending on their immediate policy or administrative needs) management or a budget. Accordingly, capacity-building often comes to be understood as a process of extending the capabilities, or increasing the size, of the entity in question. Bigger, it seems, is better, although for what end is rarely clear.

One need not look far to find the shortcomings of this idea. Take, for example, the view that equates capacity-building with budget growth. The simple appropriation of additional funds produced little or nothing in the way of desired policy outcomes in any number of cases. For example, full implementation of President George W. Bush's multibillion effort to combat AIDS in sub-Saharan Africa, generally adjudged to have made real progress, has been hampered in part because many of the recipient countries lacked the organizations or infrastructure to utilize—or in some instances, even receive—these funds (GAO 2008). Closer to the point, Marc Landy's scrutiny of EPA raises doubts as to whether a better resourced agency will be able to fulfill its multiple, often conflicting, and inextricably path-dependent missions. More resources may help counteract the agency's perceived erosion in recent years, but they will not get at EPA's core problem. J. Clarence Davies' proposed Department of Environmental and Consumer Protection (see Foreword) might be a step in the right direction because it seeks to integrate related oversight functions within a single organizational entity. On the other hand, if the critics of the mammoth Department of Homeland Security are correct, such a "meta-agency" might only produce a larger and even more unwieldy variation on a theme (May et al., 2009).

So it is important at the outset to conceptualize capacity-building as a multidimensional process. One upfront dimension of capacity is the purpose of the agency in question. What does, or will, the agency do? Government organizations engage in many different types of activities, from issuing rules to engaging in scientific research. Moreover, in order to achieve their primary goals, they must also undertake a number of subsidiary activities: recruit and train personnel, make budget requests, evaluate member performance, and so forth. Regulatory agencies in particular must have considerable capacity to deal with rapidly changing information. They must be able to scan, gather and process new information from the outside (what organization theorists know as "boundary spanning"); assimilate and integrate those inputs; and adapt successfully in response to environmental changes. In short, regulatory agencies must have the ability and opportunity to learn; acquire and integrate new knowledge and skills; accommodate new structures and procedures, and even reorganize, in response to rapid and unexpected environmental turbulence; and be resilient in responding within the context of rapid technological change.

Sometimes an organization may be called on to take on a task that is entirely new or unfamiliar. Developing this capacity may require extensive retraining or hiring of new personnel, and it might be expected to involve some—perhaps even total—reorganization. In 1958, the federal government made the unprecedented decision to initiate a civil space program. The National Aeronautics and Space Act transformed an existing organization, the National Advisory Committee on Aeronautics, into a new agency, the National Aeronautics and Space Administration (NASA). In the following years, the Eisenhower administration transferred a number of military space assets to the new civilian agency. Thus a completely new government capability was created.

This point brings us back to EPA. The agency is commonly observed to be at a crossroads 40 years after being stitched together by President Richard

Nixon out of disparate and often contradictory parts of other federal units. For one thing, a great many of its professional staff, first hired in the expansionary period of the agency's founding, are at retirement age. Who will replace them, and with what skills and orientations? Second, if one could start anew as J. Clarence Davies suggests, what would a twenty-first-century EPA look like? Even with a larger budget and rejuvenated staff capacities, would the agency be able to engage in "intra-agency innovation or the kind of agency-based policy entrepreneurship that has been so evident at the state level," as Rabe put it in Chapter 6? This is an important matter for the states, which, for all their entrepreneurship and efforts to foster mutually beneficial relationships with their industrial sectors, still depend on the federal agency and its regional offices for expert analysis and, ultimately, enforcement muscle. States like Wisconsin and California, or cities like Berkeley and Cambridge, can play essential roles in creating mechanisms for citizen input and dealing with potential local hazards from nanomaterial production or disposal, but in the end, they still rely on the federal government to fund expensive basic research, promulgate and enforce national standards, and provide a critical counterweight to the centrifugal economic and political forces that always threaten to create different classes of citizens depending on where they live.

In some ways, then, the future of nanotechnology is inextricably tied to the future of the Environmental Protection Agency—and vice versa. If, as Landy has argued, nanotechnology throws all environmental regulatory stakeholders off their accustomed games, perhaps after 40 years of existence, the time is ripe for a thorough rethinking of EPA as an institution. The promise of nanotechnology not only brings with it an imperative to rethink how we regulate, but also forces us to rethink the very organizational design of any regulatory regime.

TOWARD A TWENTY-FIRST-CENTURY REGULATORY REGIME

Government institutions are consistently portrayed as too slow and conservative to keep up with the progress of science and technology. In the twenty-first century, however, this "cultural lag" may not be so much between regulators and technology as between regulators and the public (O'Donnell et al. 2009). In particular, citizens are increasingly getting—and sharing— information on all kinds of purported environmental and health risks from any number of online sources, most of them nongovernmental. Citizens who live by Google, connect to one another through Facebook, and skim a variety of activist blogs before work every day make their decisions about new technologies and their risks well before any government regulatory agency gets around to expressing official comment. For these citizens, then, what is an appropriate and effective role for government when it is no longer seen as the primary source of critical information on environmental and health risks?

Moreover, the information available to average citizens is of such a great volume and variety that it is already running ahead of agency capacity to assimilate and use it. For example, experts at Google figured out ways to aggregate data obtained through the company's almost universally used search engines to predict flu outbreaks two weeks earlier than the Centers for Disease Control and Prevention (CDC) (Google 2009). Even the federal government is trying to get into the act, with portals such as www.regulations. gov offering means for citizen information-gathering and public comment. Even if most current users are advocacy groups of various sorts, such gateways offer citizens new vehicles for access, comment, and dissemination.

As citizen information panels like that in the city of Cambridge proliferate, and should self-reporting to internet providers become a viable alternative, regulatory agencies stand to lose ground and authority amid the noisy wealth of information. If regulatory institutions cannot learn fast enough to protect public health and welfare, or if they are seen as creating bottlenecks for innovation and economic activity, then government risks falling behind the public's own—perhaps illusory—sense that it is sufficiently informed to act on its own behalf. Coglianese (2007b, *113*) sees this possibility as real, and as a problem of efficacy and legitimacy:

> Information technology may well bring down the costs associated with accessing information and submitting comments to agencies, but many rules will continue to have significant consequences for citizens without eliciting much public attention. Other barriers to citizen participation will remain, perhaps most saliently, the specialized knowledge requisite to meaningful participation—not only familiarity with the organization and operation of government but with the technical issues underlying a given rulemaking. After all, if the issues underlying rulemaking were sufficiently technical or difficult as to lead Congress to delegate to an expert agency to begin with, then by definition these issues will be difficult for ordinary citizens to understand. Moreover, even with greater accessibility to rulemaking information via the Internet, most citizens are unlikely to have or to take the time to learn about the technical issues surrounding rulemaking.

Regulatory agencies, for all their efforts to foster greater citizen access to online resources, will need to be even more serious about the value of transparency and responsiveness in an information-infused world. Transparency can bolster scientific and agency credibility as it engages the public in conversations about risk, uncertainty, and reward. Ultimately, government is uniquely situated to facilitate communication among a broad range of stakeholders, including citizens, adding to the possibility that long-standing misunderstanding and fears that have factionalized regulatory politics in the past may be openly addressed and redressed. If nothing else, greater transparency and public engagement can lead us away from the limited discourse of technocratic elites about public fear (and fear of the public) and, instead, provide more direct access to understanding what fears the public, scientists, and industry actually share. By focusing on transparency, credibility, and public engagement, agencies have the practical means to garner public trust. Trust earned in this way is transitive, providing the public with reasons to trust in both the innovations derived through nanotechnology and the judgments of its twenty-first-century regulatory agencies.

Thus a large part of the near-term challenge is for regulatory agencies to realize that nanotechnology in and of itself, combined with powerful transformations in information access and exchange, will challenge the basic orientation of environmental regulation. To put government in the service of the broad public good does not mean a return to command and control models of regulatory governance. The nature of nanotechnology and the global marketplace within which it emerges probably militates against such an approach. Rather, an effective and legitimate twenty-first-century regulatory regime will be one that enlists many possible actors in a transparent public discourse whose ultimate aim is to reap the benefits of technology development while also inserting broader social norms and goals into the discourse. The dreams of democratic governance, arguably equally important to any dreams of nanotechnology, demand such engagement. As Rabe asserts in Chapter 6, doing this "is likely to necessitate a type of governance that defies conventional hierarchical patterns." Indeed, all of the skills embedded in network approaches to governance will likely prove essential (Goldsmith and Eggers 2004). Such networks will be inherently messy, but in the end, they will produce more informed, fairer, and more legitimate outcomes.

To echo Landy in Chapter 5, a new path dependency on environmental governance is being created in real time as we write. The policy decisions made in the earliest days of nanotechnology will resonate for decades to come. Getting it right will matter.

REFERENCES

Badash, Lawrence, Elizabeth Hodes, and Adolph Tiddens. 1986. Nuclear Fission: Reaction to the Discovery in 1939. *Proceedings of the American Philosophical Society* 130 (2): 196–231.

Beierle, Thomas C., and Jerry Cayford. 2002. *Democracy in Practice: Public Participation in Environmental Decisions*. Washington, DC: Resources for the Future.

Berube, David M. 2006. *Nano-Hype: The Truth behind the Nanotechnology Buzz*. Amherst, NY: Prometheus Books.

Bosso, Christopher, and Ruben Rodrigues. 2007. Organizing around Emerging Issues: Interest Groups and the Making of Nanotechnology Policy. In *Interest Group Politics*, 7th ed., edited by Allan Cigler and Burdett Loomis. Washington, DC: CQ Press, 366–88.

Coglianese, Cary. 2007a. Business Interests and Information in Environmental Rulemaking. In *Business and Environmental Policy*, edited by Michael Kraft and Sheldon Kamieniecki. Cambridge, MA: MIT Press, 185–210.

———. 2007b. Weak Democracy, Strong Information: The Role of Information Technology in the Rulemaking Process. In *Governance and Information Technology: From Electronic Government to Information Government*, edited by Victor Mayer-Schonberger and David Lazer. Cambridge, MA: MIT Press, 101–122.

Davies, J. Clarence. 2009. Revising the Toxic Substances Control Act of 1976. Testimony given to the Subcommittee on Commerce, Trade, and Consumer Protection, of the Committee on Energy and Commerce, U.S. House of Representatives, 111th Cong., 1st sess., February 26. energycommerce.house.gov/Press_111/20090226/testimony_davies.pdf (accessed March 20, 2009).

Dyson, George. 2002. *Project Orion: The True Story of the Atomic Spaceship*. New York, NY: MacMillan.

Frederickson, David G., and H. George Frederickson. 2006. *Measuring the Performance of the Hollow State*. Washington, DC: Georgetown University Press.

GAO (U.S. Government Accountability Office). 2008. *Global HIV/AIDS: A More Country-Based Approach Could Improve Allocation of PEPFAR Funding*. Report #GAO-08-480 (April). www.gao.gov/new.items/d08480.pdf (accessed March 20, 2009).

Goldsmith, Stephen, and William D. Eggers. 2004. *Governing by Network: The New Shape of the Public Sector*. Washington, DC: Brookings Institution.

Google. 2009. Flu Trends. www.google.org/about/flutrends/how.html (accessed April 4, 2009).

Kay, W. D., and Hans Eijmberts. 2009. The Perils of Emerging Technologies: What the Nanotechnology Community Can Learn from the Experience of Nuclear Physics. Paper presented at the annual meeting of the Midwest Political Science Association. April 3–4, 2009, Chicago, IL.

Kettl, Donald F., ed. 2002. *Environmental Governance: A Report on the Next Generation of Environmental Policy*. Washington, DC: Brookings Institution.

Kulinowski, Kristen. 2004. Nanotechnology: From Wow to Yuck? *Bulletin of Science, Technology & Society* 24 (February): 13–20.

Lipson, Sam. 2003. The Cambridge Model: How Public Oversight of Biotech Is Good for Everyone—Even Business. *Gene Watch* 16 (September–October): 7–10.

May, Peter J., Ashley E. Jochim, and Joshua Sapotichne. 2009. Policy Regimes and Governance: Constructing Homeland Security. Paper presented at the 10th Public Management Research Association Conference, October 2009. www.pmranet. org/conferences/OSU2009/papers.html (accessed December 4, 2009).

Meins, Erika. 2003. *Politics and Public Outrage: Explaining Transatlantic and Intra-European Diversity of Regulations on Food Irradiation and Genetically Modified Food.* Münster, Germany: LIT Verlag.

O'Donnell, Sean T., Ronald Sandler, and Christopher Bosso. 2009. Emerging Technologies: Nanotechnology and Regulatory Regimes. *The Nanotechnology Food, Drug, Cosmetic and Medical Device Regulatory Guide.* Washington, DC: Food and Drug and Law Institute, 195–204.

Weart, Spencer R. 1988. *Nuclear Fear: A History of Images.* Cambridge, MA: Harvard University Press.

Weisskopf, Victor F. 1970. Physics in the Twentieth Century. *Science* 168 (3934): 923–30.

White House. 2000. *National Nanotechnology Initiative: Leading to the Next Industrial Revolution.* Washington, DC: White House, Office of the Press Secretary. clinton4. nara.gov/WH/New/html/20000121_4.html (accessed March 19, 2009).

INDEX